W0071649

ANALYTICAL TECHNIQUES FOR NATURAL PRODUCT RESEARCH

FSC
www.fsc.org
MIX
Paper from
responsible sources
FSC® C013604

ANALYTICAL TECHNIQUES FOR NATURAL PRODUCT RESEARCH

Satyanshu Kumar

www.cabi.org

CABI is a trading name of CAB International

CABI	CABI
Nosworthy Way	745 Atlantic Avenue
Wallingford	8th Floor
Oxfordshire OX10 8DE	Boston, MA 02111
UK	USA
Tel: +44 (0)1491 832111	Tel: +1 (617) 682 9015
Fax: +44 (0)1491 833508	E-mail: cabi-nao@cabi.org
E-mail: info@cabi.org	
Website: www.cabi.org	

© Satyanshu Kumar 2016. All rights reserved. No part of this publication may be reproduced in any form or by any means, electronically, mechanically, by photocopying, recording or otherwise, without the prior permission of the copyright owners.

A catalogue record for this book is available from the British Library, London, UK.

Library of Congress Cataloging-in-Publication Data

Names: Kumar, Satyanshu, author.
Title: Analytical techniques for natural product research / Satyanshu Kumar.
Description: Boston, MA : CABI, 2015.
Identifiers: LCCN 2015031745 | ISBN 9781780644738 (alk. paper)
Subjects: LCSH: Plant extracts--Analysis--Technique.
Classification: LCC QK861 .K86 2015 | DDC 615.3/21--dc23 LC record available at http://lccn.loc.gov/2015031745

ISBN-13: 978 1 78064 473 8

Commissioning editor: Nicki Dennis
Editorial assistant: Emma McCann
Production editor: Tracy Head

Typeset by SPi, Pondicherry, India.
Printed and bound in the UK by CPI Group (UK) Ltd, Croydon, CR0 4YY.

Contents

Preface

Natural products, including plants, animals, marine life and minerals, are a huge reservoir of chemical diversity. Ancient wisdom and knowledge have formed the basis of modern medicine. The development of plant-derived drugs based on the leads obtained from traditional systems of medicine has been the thrust of drug discovery. Plant-derived, biologically active molecules have played a significant role in human health care. The World Health Organization has been active in creating strategies, guidelines and standards for plant-based medicines. Accordingly, many international authorities and agencies have started creating new mechanisms to induce and regulate quality control and standardization of plant-based medicines. Advance analytical techniques, hyphenated techniques in particular, have made it easier to standardize multi-component, plant-based formulations. High-throughput techniques are being used for screening a large number of extracts, enriched fractions, pure molecules, etc., in the itinerary of drug discovery. Furthermore, with the advancement of modern science, the need for a quality standard for plant-based drugs has also been stressed and regulations are coming into force across the globe.

Knowledge on such aspects is essential for creating awareness to increase the use efficiency of raw plant materials, as well as their processing. Therefore, it was thought prudent to collate and compile information describing different analytical techniques. It is expected that the present publication will arouse great interest among readers from natural product research and related subjects.

It is my duty to express sincere thanks to Dr Sreepat Jain, formerly Commissioning Editor, CABI, New Delhi, Nicki Dennis, Commissioning Editor, Tracy Head, Senior Production Editor, Emma McCann, Editorial Assistant and the CABI team for their constant support for completing this manuscript. I also express heartfelt thanks to my wife Seema, my sons Pranay and Yash, and my friends and colleagues at the ICAR-Directorate of Medicinal and Aromatic Plants Research, Boriavi, Anand, Gujarat, India for their support and encouragement.

1 Analytical Techniques in Natural Product Research

1.1 Introduction

Nature represents an extraordinary reservoir of novel molecules. Natural products have provided the inspiration for most of the active ingredients in medicines. Their high chemical diversity and the effects of evolutionary pressure to create biologically active molecules could be attributed to the success in drug discovery. Medicinal plants have played a key role in world health. Plants are rich sources of fine chemicals, largely unknown and explored, yet they still make an important contribution to health care in spite of the great advances in the field of modern medicine. Plants are a treasure trove of interesting and valuable compounds since they must glean everything from the spot on the earth where they are rooted. Also, they cannot escape when threatened; therefore, they have evolved a most impressive panoply of products to thrive in ever-changing environments despite these limitations (Newell-McGloughlin, 2008). There are about 400,000 higher plant species in the world (Hostettmann and Terreaux, 2000). It is estimated that plants produce up to 200,000 phytochemicals across their many and diverse members (Oksman-Caldentey and Inze, 2004). It is estimated that about 25% of all modern medicines are directly or indirectly derived from higher plants (Fransworth and Morris, 1976; Cragg *et al.*, 1997; De Smet, 1997; Shu, 1998). Many of these pharmaceuticals are still in use today and often no useful synthetic substitutes have been found that possess the same efficacy and pharmacological specificity to a particular disease. In some cases, such as for antitumoral and antimicrobial drugs, about 60% of the current available medicines and others in the late stages of clinical trials have been derived from natural products, mainly from higher plants (Cragg *et al.*, 1997). Plant-based drugs and formulations have been used since ancient times as a remedy for a range of diseases. About 65–80% of the world's population from developing countries essentially depends on plants for primary health care (Akerele, 1993). Besides plants, marine natural products and microorganisms are also a major source for new drugs (Planes and Caballero-George, 2015). Oceans encompass a stressful and competitive habitat with unique conditions of pH, temperature, pressure, oxygen, light, nutrients and salinity. These factors force organisms to adapt both chemically and physiologically to survive (Tsurumi *et al.*, 1995; Skropeta and Wei, 2014).

Marine natural products are potential sources for pharmaceuticals such as antimicrobial, antiviral, antiparasitic, anticancer, anti-inflammatory, neuroprotective and immunomodulatory agents (Konig and Wright, 1996; Mayer and Hamann, 2002; Abad *et al.*, 2008; Mayer *et al.*, 2013; Blunt *et al.*, 2014; Skropeta and Wei, 2014). More than 50,000 microbial natural products also have an important role in drug discovery (Xiong *et al.*, 2013). A renewed interest in investigating higher plants as sources for new lead structures and also for the development of standardized phytotherapeutic agents with proven efficacy, safety and quality has been demonstrated (Brevoort, 1995; De Smet, 1997; Blumenthal, 1999a,b). More than 20 new drugs that were launched globally between 2000 and 2005 originated from natural products (Tasduq *et al.*, 2008).

1.2   Role of Secondary Metabolites

Natural product chemistry has evolved into an interdisciplinary area of science concerned with the isolation, characterization and determination of biological activity of pure phytochemicals, as well as extracts or enriched fractions. Phytochemicals, the active components for biological activity, are generally referred to as secondary metabolites. Plant secondary metabolites are low molecular weight compounds and can be defined as compounds that have no recognized role in the maintenance of fundamental life processes in the plants that synthesize them, but they do have an important role in the interaction of the plant with its environment (Oksman-Caldentey and Inze, 2004). They are characterized by structural diversity and are synthesized from a limited pool of biosynthetic precursors: phosphoenolpyruvate, pyruvate, acetate, amino acids, acetyl CoA and malonyl CoA. Although various hypotheses have been proposed to account for the production of secondary metabolites, none is entirely satisfactory. However, information on their biosynthesis is essential to understand the interaction between plants and the environment (Robards *et al.*, 1999).

The production of these compounds is often low (less than 1% of dry weight) and depends on the physiological and developmental stage of the plant. Secondary metabolites are characterized by wide chemical diversity, and every plant has its own characteristic set of secondary metabolites. Based on their biosynthetic origins, plant secondary metabolites can be divided structurally into five major groups: polyketides, isoprenoids, alkaloids, phenylpropanoids and flavonoids. It is generally accepted that plant secondary metabolites are important for the survival of the plant, and in its ecosystem their antimicrobial and anti-insect activities deter potential predators, discourage competing plant species and attract pollinators or symbionts (Dixon, 2001). Secondary metabolites have for centuries been of interest to humans as flavours, fragrances, dyes, pesticides and pharmaceuticals, as well as being important for plant itself.

1.3 Natural Product Research and Analytical Techniques

Traditional medicine and herbal products are generally plant derived and consist of hundreds of unknown components rather than a single component or a simple combination of several components. Also, many of the components are low in quantity. Usually, the active principles responsible for the pharmacological action are unknown. Multiple active components, including macro and micro components, are frequently considered to be responsible for the therapeutic effects, and thus the analysis of multiple components is more reasonable for quality control. Furthermore, herbal drugs, individually and in combination, contain a myriad of compounds in complex matrices in which no single active constituent is responsible for overall efficacy. Consequently, simultaneous quantitative analysis of various kinds of active components is the most direct and important method for quality control. Despite the availability of modern analytical instrumentation techniques (Fig. 1.1), rarely

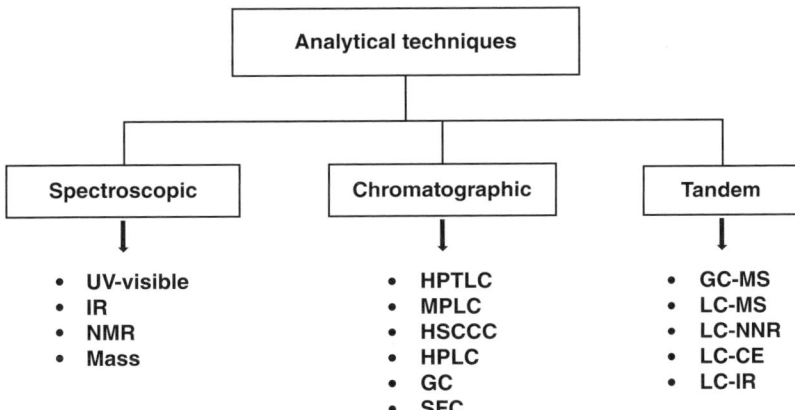

Fig. 1.1. Schematic diagram of analytical techniques used in natural product research. IR = infrared; NMR = nuclear magnetic resonance; HPTLC = high-performance thin-layer chromatography; MPLC = medium-pressure liquid chromatography; HSCCC = high-speed countercurrent chromatography; HPLC = high-performance liquid chromatography; GC = gas chromatography; SFC = supercritical fluid chromatography; GC-MS = gas chromatography–mass spectrometry; LC–MS = liquid chromatography–mass spectrometry; LC-NMR = liquid chromatography–nuclear magnetic resonance; LC-CE = liquid chromatography–capillary electrophoresis; LC-IR = liquid chromatography–infrared.

do phytochemical investigations succeed in isolating and characterizing all secondary metabolites present in the plant extract (Calixto, 2000). Furthermore, only about 10% of higher plant species have been characterized chemically to some extent. Chemical complexity makes the quality control process much more complicated. However, it has become inherent to determine the chemical profile of plant-based products for better scientific and clinical acceptability, as well as for proper global positioning. Despite their existence and continued use over many centuries and their popularity and extensive use during the past decades, traditional medicines have not been officially recognized in most countries. The quantity and quality of safety and efficacy data on traditional medicines are far from sufficient to meet the criteria needed to support their use worldwide (WHO, 2000). Furthermore, the chemical constituents in component plants may vary depending on harvest seasons, plant origin and other postharvest processes. To ensure the reliability and repeatability of pharmacological and clinical research, as well as also to understand their bioactivities and possible side effects, it is essential to determine most of the phytochemical constituents of traditional medicines (Bauer, 1998; Raven *et al.*, 1999; Yan *et al.*, 1999). It is also well known that the efficacy of traditional medicines has a characteristic of a complex mixture of chemical compounds present in the crude drug; reasonable evaluation of their relationship is not a trivial task. A chemical fingerprint can be linked to biological assays to provide assurance of efficacy and consistency. However, the research work on this aspect is far from sufficient to meet the criteria needed.

   Isolation of the ingredients of plant extracts in adequate quantities for spectral and biological assays is the basis of phytochemical research. Rapid identification and quantification of biologically active natural products plays a strategic role in phytochemical investigations of crude plant extracts. Also, the dereplication of crude extracts prior to isolation work is crucial to avoid the tedious isolation of known constituents. Recent advances in the area of the purification process, isolation and structure elucidation have made it possible to establish appropriate strategies for the quality control and standardization of herbal products in order to maintain as much as possible the homogeneity of the plant extract, and therefore the derived product or formulation. A wide spectrum of analytical methods has been developed for application in phytochemical research, pharmacological studies or quality control. The analytical approaches can be classified as: (i) the analysis of targeted compounds, i.e. specifically one compound; or (ii) group-specific analysis, i.e. the analysis of a number of (preferably all) compounds belonging to that particular group, and metabolite profiling aimed at a large number of primary and secondary metabolites from the extract, including carbohydrates, lipids, amino acids, etc. Metabolomics involves the analysis of the entire metabolome as the sum of all detectable low and intermediate molecular mass compounds in place of individual (target) metabolites. Different approaches such as metabolite profile and metabolic fingerprinting are followed in the field of metabolomics (Goodacre *et al.*, 2004). Metabolic profiling covers the identification of a selected group of metabolites, whereas metabolic fingerprinting comprises the classification of samples on the basis of provenance of either biological relevance or origin (Angelova *et al.*, 2008). Metabolomics has developed into an important tool for applications in natural product studies and quality control, as well as in studies on diseases and toxicity (Verpoorte *et al.*, 2007).

Traditionally, histological and morphological inspections have been the usual methods of authentication. But these methods cannot be applied to the final forms of modern herbal products such as herbal extracts and dosage forms. Chemical and chromatographic techniques are currently used for the identification and assessment of components (Angelova *et al.*, 2008). Phytochemical analysis is commonly performed using standard techniques such as thin-layer chromatography (TLC), high-performance thin-layer chromatography (HPTLC), high-performance liquid chromatography (HPLC), gas chromatography (GC) and, more recently, mass spectrometry (MS) and nuclear magnetic resonance (NMR) spectrometry. Significant improvements have been made in the separation and resolution of analytical techniques. High-resolution instruments are now becoming available to a broader group of researchers. Furthermore, with the developments in biostatics with the aid of multivariate analysis (pattern recognition), it is possible to compare the metabolite profile, and the differences between a group of samples can be identified rapidly, thereby rendering the identification and quantification of all individual metabolites present in the profiles superfluous. The holistic approach requires a quite different procedure than traditional analysis, as instead of focusing on a few selected compounds, it is imperative to generate a profile of a large number of constituents, aiming at the identification of novel biomarkers and drug targets. NMR and liquid chromatography–mass spectrometry (LC-MS) are the most popular techniques for metabolite profiling and fingerprinting. NMR spectra of unpurified solvent plant extracts have the potential to provide relatively unbiased fingerprints, comprising overlapping signals of the majority of the metabolites present in the solution (Ward *et al.*, 2002). An NMR-based study on the influence of the composition of the extraction solvent in relation to the quality of the metabolite profile has also been reported (Angelova *et al.*, 2008). Two-dimensional *J*-resolved NMR spectra and multivariate data analysis techniques were applied in order to avoid low resolution and overlapping signals hampering the identification of the individual components of ginseng roots (Yang *et al.*, 2006). The elemental composition of metabolites is one of the most valuable pieces of information for identification purposes. Accurate mass measurements, which means by definition that the measured mass should be within 5 ppm of the theoretical mass, can be obtained from the latest generation Fourier transform ion cyclotron resonance (FT-ICR), thereby allowing unequivocal determination of the elemental composition. These accurate mass data are obtainable on a chromatographic timescale and without the need for internal calibration (Van der Greef *et al.*, 2004).

For a detailed analysis of the metabolome, chromatographic procedures are often preferred. Although there are many chromatographic techniques including hyphenated chromatography available for instrumental analysis of natural products, TLC was the common method of choice before instrumental chromatography methods like GC and HPLC were established. Even nowadays, TLC is still frequently used for the analysis of herbal products as an easier method of preliminary screening with a semi-quantitative evaluation together with other chromatographic techniques. Simplicity, versatility, high velocity, specific sensitivity and simple sample preparation are the advantages of using TLC for constructing fingerprints. Various pharmacopoeias such as the *American Herbal Pharmacopoeia, Chinese Drug Monographs and Analysis, Pharmacopoeia of the People's Republic of China*, etc., still permit the use of TLC for providing the first characteristic fingerprints of herbs. TLC has advantages of manyfold possibilities of detection in the analysis of herbal products. Further, with the help of a video store system, it is possible to gather useful qualitative and quantitative information from the developed TLC plate (Chau *et al.*, 1998). TLC is also in the process of being updated. Forced-flow planar chromatography (FFPC) uses hydrostatic pressure to increase the velocity of the mobile phase. Rotation planar chromatography (RPC), overpressured layer chromatography (OPLC) and electro planar chromatography (EPC) are the other innovations. Parallel and serially coupled layers open up new avenues for the analysis of a large number of samples (up to 216) for high-throughput screening and for the analysis of very complex matrices (Nyiredy, 2003).

The methods of extraction and sample preparation are also of great importance in preparing good fingerprints. The main goal of extraction is to obtain the maximum number of metabolites or, ideally, all the metabolites present in the samples. Different kinds of extraction methods are usually used applying different solvent combinations. Soxhlet extraction is one of the oldest methods for solid–liquid extraction. It has been regarded as the reference for the optimization of new extraction techniques and has been the most cited among the other extraction techniques used. Ultrasound-assisted solvent extraction, microwave-assisted extraction and supercritical fluid extraction methods have emerged as modern extraction techniques. Solid-phase extractions, besides being used for sample clean up or concentration of metabolites from a liquid matrix, are also utilized for extraction purposes. This technique has been automated and coupled with online extraction and analytical instruments such as GC or LC-MS (Louter *et al.*, 1999; Lopez-Blanco *et al.*, 2002).

The concept of phytoequivalence was developed in Germany to ensure the consistency of herbal products (Tyler, 1999). According to this concept, a chemical profile such as a chromatographic fingerprint for a herbal product should be constructed and compared with the profile of a clinically proven reference product, as it is almost impossible to develop an appropriate analytical method (including sample preparation and chromatographic procedures) to represent all the constituents of the chemical characteristics in a chromatogram (Fan *et al.*, 2006). Therefore, the development of multiple chromatographic fingerprints has been suggested. Advances in hyphenation techniques in chromatographic instruments can make the quality control of natural products, both in qualitative and quantitative regards, stronger and stronger. High-throughput analysis and miniaturization that provide high resolution in a short analysis time with low operational costs are desirable. The introduction of sub-2-µm columns for HPLC and the development of ultra-high pressure liquid chromatography (UHPLC) represent important steps forward for crude extract profiling. The speed of separation provided by these new separation methods challenges the liquid chromatography detectors that have to provide faster responses. The strong development of '-omics' (metabolomics, genomics, etc.) over the last decade also challenges the analytical methods as they aim to detect and quantify all metabolites in a given organism. For these types of studies, both separative and non-separative methods must be complementary because none of them alone can fulfil all the ideal needs of sensitivity and throughput (Wolfender, 2008).

1.4 Forms of Natural Products and Good Manufacturing Practices

There are several common forms of natural products, including phytochemicals, nutraceuticals, cosmeceuticals, oleoresins, essential oils, etc. Phytochemicals refers to the chemicals present in plants. Nutraceuticals are any substance that may be considered as a food or a part of food providing health benefits, including the prevention and treatment of diseases. Oleoresins are pure extractives derived from spices containing concentrated natural flavouring components both volatile and non-volatile. Cosmeceuticals is a term to describe cosmetic-containing ingredients. The volatile part of the plant largely responsible for its characteristic aroma comes under the category of essential oil.

Good manufacturing practice (GMP) is a code of practice used for maintaining the highest standard of quality in the process of the production and control of natural products, in particular in herbal products.

1.5 Conclusion

The fingerprinting technique has been widely accepted as a useful method for the evaluation and quality control of raw materials and their finished products from natural sources. Analysing chemical markers that are known to be present in the natural product is one of the common quality control methods used in research laboratories in industry. Variances due to geographical source, cultivation and processing methods affect chemical composition and clinical efficacy. Therefore, it is necessary to establish a method to control the quality. Furthermore, multi-sourcing has been a major cause of clinical accidents in phytotherapies. Physically similar plants from the same or even different genera are used as the same herb. The differences in the chemical compositions of various species may lead to different biological activities (Ye *et al.*, 2007). HPLC is efficient in separating the chemical compounds in a mixture and, using mass spectrometry, sufficient information for structural elucidation of the compounds could be generated when tandem mass spectrometry (MS^n) is applied. Therefore, the combination of HPLC and MS facilitates the rapid and accurate identification of chemical compounds, especially when a pure standard is unavailable. The identification of the compound is confirmed using ultraviolet spectroscopy (UV), infrared spectroscopy (IR), nuclear magnetic resonance spectroscopy (NMR) and mass spectroscopy (MS).

References

Abad, M.J., Bedoya, L.M. and Bermejo, P. (2008) Natural marine anti-inflammatory products. *Mini Review in Medicinal Chemistry* 8, 740–754.

Akerele, O. (1993) Summary of WHO guidelines for the assessment of herbal medicines. *Herbal Gram* 28, 13–19.

Angelova, N., Kong, H.W., Heijden, R.V., Yang, S.Y., Choi, Y.H., *et al.* (2008) Recent methodology in the phytochemical analysis of Ginseng. *Phytochemical Analysis* 19, 2–16.

Bauer, R. (1998) Quality criteria and standardization of phytopharmaceuticals: can acceptable drug standards be achieved? *Drug Information Journal* 32, 101–110.

Blunt, J.W., Copp, B.R., Keyzers, R.A., Munro, M.H. and Prinsep, M.R. (2014) Marine natural products. *Natural Products Reports* 31, 160–258.

Blumenthal, M. (1999a) Havard study estimates consumers spend $5.1 billion on herbal products? *Herbal Gram* 45, 68.

Blumenthal, M. (1999b) Herb industry sees merges, acquisition and entry by pharmaceutical giants in 1988. *Herbal Gram* 45, 67–68.

Brevoort, P. (1995) The US botanical market. An overview. *Herbal Gram* 36, 49–59.

Calixto, J. (2000) Efficacy, safety, quality control, marketing and regulatory guidelines for herbal medicines (phytotherapeutic agents). *Brazilian Journal of Medical and Biological Research* 33, 179–189.

Chau, F.T., Chan, T.P. and Wang, J. (1998) TLCQA: quantitative study of thin-layer chromatography. *Bioinformatics* 14, 540–541.

Cragg, G.M., Newman, D.G. and Snader, K.M. (1997) Natural products in drug discovery and development. *Journal of Natural Products* 60, 52–60.

De Smet, P.A.G.M. (1997) The role of plant derived drugs and herbal medicines in healthcare. *Drugs* 54, 801–840.

Dixon, R.A. (2001) Natural products and plant disease resistance. *Nature* 411, 843–847.

Fan, X.-H., Cheng, Y.-Y., Ye, Z.-L., Lin, R.-C. and Qian, Z.-Z. (2006) Multiple chromatographic fingerprinting and its application to the quality control of herbal medicines. *Analytica Chimica Acta* 555, 217–224.

Fransworth, N.R. and Morris, R.W. (1976) Higher plants – the sleeping giant of drug development. *American Journal of Pharmaceutical Education* 148, 46–52.

Goodacre, R., Vaidyanathan, S., Dunn, W.B., Harrigan, G. and Kell, D.B. (2004) Metabolomics by numbers: acquiring and understanding global metabolite data. *Trends in Biotechnology* 22, 245–252.

Hostettmann, K. and Terreaux, C. (2000) Search for new lead compounds from higher plants. *Chimia (Aarau)* 54, 652–657.

Konig, G. and Wright, A.D. (1996) Marine natural products research: current directions and future potential. *Planta Medica* 62, 193–211.

Lopez-Blanco, M.C., Reboreda-Rodriguez, B., Cancho-Grande, B. and Simal Gandara, J. (2002) Optimization of solid-phase extraction and solid-phase microextraction for the determination of alpha- and beta-endosulfan in water by gas chromatography-electron-capture detection. *Journal of Chromatography A* 976, 293–299.

Louter, A.J.H., Vreuls, J.J. and Brinkman, U.A.T. (1999) On-line combination of aqueous sample preparation and capillary gas chromatography. *Journal of Chromatography A* 842, 391–426.

Mayer, A.M. and Hamann, M.T. (2002) Marine pharmacology in 1999: compounds with anti-bacterial, anticoagulant, antifungal, anthelmintic, anti-inflammatory, antiplatelet, antiproto-zoal and antiviral activities affecting the cardiovascular endocrine, immune and nervous systems and other miscellaneous mechanisms of action. *Comparative Biochemistry and Physiology – Part C: Toxicology & Pharmacology* 132, 315–339.

Mayer, A.M., Rodriguez, A.D., Taglialatela-Scafati, O. and Fusetani, N. (2013) Marine pharmacology in 2009–2011: marine compounds with antibacterial, antidiabetic, antifungal, anti-inflammatory, anti-protozoal, antituberculosis and anti-viral activities: affecting the immune and nervous systems and other miscellaneous mechanisms of action. *Marine Drugs* 1, 2510–2573.

Newell-McGloughlin, M. (2008) Nutritionally improved agricultural crops. *Plant Physiology* 147, 939–953.

Nyiredy, S. (2003) Progress in forced-flow planar chromatography. *Journal of Chromatography A* 1000, 985–999.

Oksman-Caldentey, K.M. and Inze, D. (2004) Plant cell factories in the post-genomic era: new ways to produce designer secondary metabolites. *Trends in Plant Science* 9, 433–440.

Planes, N. and Caballero-George, C. (2015) Marine and soil derived natural products: a new source of novel cardiovascular protective agents targeting the endothelin system. *Planta Medica* 81, 630–636.

Raven, P.H., Evert, R.F. and Eichhorn, S.E. (1999) *Biology of Plants*, 6th edn. Freeman, New York.

Robards, K., Prenzler, P.D., Tucker, G., Swatsitang, P. and Glover, W. (1999) Phenolic com-pounds and their role in oxidative processes in fruits. *Food Chemistry* 66, 401–436.

Shu, Y.Z. (1998) Recent natural products based drug development: a pharmaceutical industry perspective. *Journal of Natural Products* 61, 1053–1071.

Skropeta, D. and Wei, L. (2014) Recent advances in deep sea natural products. *Natural Products Reports* 31, 999–1025.

Tasduq, S.A., Kaiser, P.J., Gupta, B.D., Gupta, V.K. and Johri, R.K. (2008) Negundoside, an iridiod glycoside from leaves of *Vitex negundo*, protects human liver cell against calcium-mediated toxicity induced by carbon tetrachloride. *World Journal of Gastroenterology* 21, 3693–3709.

Tsurumi, Y., Fujie, K., Nishikawa, M., Kiyoto, S. and Okuhara, M. (1995) Biological and pharma-cological properties of highly selective new endothelin converting enzyme inhibitor WS79089B isolated from *Streptosporangium roseum* No. 79089. *Journal of Antibiotics (Tokyo)* 48, 169–174.

Tyler, V.E. (1999) Phytomedicines: back to the future. *Journal of Natural Products* 62, 1589–1592.

Van der Greef, J., Van der Heijden, R. and Verheij, E.R. (2004) The role of mass spectrometry in system biology: data processing and identification strategies in metabolomics. In: Ashcroft, A.E., Brenton, G. and Monaghan, J.J. (eds) *Advances in Mass Spectrometry*, Vol. 16. Elsevier Science, Amsterdam, pp. 145–164.

Verpoorte, R., Choi, Y.H. and Kim, H.K. (2007) NMR based metabolomics at work in phyto-chemistry. *Phytochemistry Reviews* 6, 3–14.

Ward, J.L., Harris, C., Lewis, J. and Beale, M.H. (2002) Assessment of ^{1}H NMR spectros-copy and multivariate analysis as a technique for metabolite fingerprinting of *Arabidopsis thaliana. Phytochemistry* 62, 949–957.

WHO (2000) General guidelines for methodologies on research and evaluation of traditional medicine. World Health Organization, Geneva, Switzerland, 80 pp.

Wolfender, J.L. (2008) HPLC in natural product analysis: the detection issue. *Planta Medica* 75, 719–734.

Xiong, Z., Wang, J., Hao, J. and Wang, Y. (2013) Recent advances in the discovery and devel-opment of marine natural products. *Marine Drugs* 11, 700–717.

Yan, X.J., Zhou, J.J., Xie, G.R. and Milne, G.W.A. (1999) *Traditional Chinese Medicines: Molecular Structures, Natural Sources and Applications.* Ashgate, Aldershot, UK.

Yang, S.Y., Kim, H.K., Lefeber, A.W.M., Erkelens, C., Angelova, N., *et al.* (2006) Application of two-dimensional nuclear magnetic resonance spectroscopy to quality control of ginseng commercial products. *Planta Medica* 72, 364–369.

Ye, M., Han, J., Chen, H., Zheng, J. and Guo, D. (2007) Analysis of phenolic compounds in rhu-barbs using liquid chromatography coupled with electrospray ionization mass spectrometry. *Journal of American Society of Mass Spectrometry* 18, 82–91.

2 Phytochemical Processing: Extraction Methods

2.1 Introduction

The plant kingdom represents an enormous reservoir of biologically active molecules, and plants with ethnopharmacological information have been the main source of early drug discovery. A large proportion of the drugs used in modern medicine have either been discovered directly from plants or modified synthetically from a lead compound. In addition, in the form of natural products or as functional foods, plants and their extracts offer an alternative to specifically targeted drugs in the treatment and prevention of many diseases. As diverse biological activities of plant extracts and phytochemicals are being reported, investigations of higher plants with known ethnobotanical information have attracted the attention of researchers. Phytochemicals have high commercial value in the local and global markets because of shifting from illness-oriented products to wellness-promoting products. Besides that, the prevalence of chronic diseases that cannot be cured by conventional drugs makes the phytochemical industry an upcoming industrial sector. However, a common pitfall associated with this sector is that the production of these phytochemicals is carried out mainly through various traditional methods, leading to high losses and low yield. To make the phytochemical industries viable and profitable, various transformations like suitable processes that include planting and harvesting, raw material preparation and value-added production are needed. Also, for the successful modernization of phytochemical processing, process technology needs to be optimized for extraction and product formulation. Several steps involved in phytochemical processing include size reduction through to chopping and grinding to good storage, to ensure that active phytochemicals are maintained before processing. Extraction is a key step in the iterative process of drug discovery from the lead discovery stage. Extracts of medicinal plants are the intermediate stage between the original plant and isolated activity: plant-based formulations should be as close to the original plant as active components, in order to reduce any possibility of deterioration in therapeutic values (Qu *et al.*, 2006). However, in practice, physical and chemical pretreatments are required to remove unwanted impurities that can influence the pharmacodynamic effects of the active components. Extraction is followed by isolation of the lead compounds, coupled with improvement by the rational design and synthesis of new analogues with better pharmacological profiles. For phytochemical analysis also, this is a very important step. The extract obtained is standardized in terms of chemical and active markers. Chemical markers are the major constituents of the plant, whereas active markers are the chemical constituents responsible for bioactive properties. Both types of markers are utilized for establishing the relationship between classes of plants and the occurrence of specific groups in chemosystematic studies.

Crude plant extract is first assayed for particular bioactivities based on the available ethnopharmacological information and this is followed by preparation of the extracts and bioactivity guided isolation of marker compounds. Extraction technology is optimized in terms of operating parameters as well as for yield of desired constituents.

Standardization of the processing method is important for increasing the value of the extracts and is critical, especially due to low yield of extracts. It also ensures product potency through consistency in the content of active compounds. With increased processing, the value of the product is also increased; however, in some cases, it has been reported that the extract is more effective than the pure phytochemicals, possibly due to the synergistic action of phytochemicals. Therefore, standardization of the extract must assure that the phytochemical profile of the plant is maintained.

Although primary emphasis in phytochemical processing is on improving processing techniques and the developing products, optimization of the extraction process, which includes yield enhancement, reduction in the volume of solvents used for the extraction and reduction in processing time, is also of importance. Biorefinery is a new concept where raw plant material is totally fractionated and converted into a spectrum of valuable products. This concept was applied to cinnamon, a common spice used in Asian markets, which includes essential oil, oleoresins, extracts and purified phytochemicals. Methyl hydroxyl chalcone polymer (MHCP) is one high-value product from cinnamon with increased cellular oxidation capacity, improved functional capacity of insulin receptors in cells and a strong antioxidant capacity (Anderson and Schmidt, 2000). An environmentally safe and economically viable production technology would be useful for extracts as functional food additives or phytomedicines.

2.2 Steps before Extraction from Plant Materials

2.2.1 Collection of plant materials: GAP and GACP protocols

Collection of plant materials, whole plant, aerial part and tissue (leaves, stem, bark and roots), is the first step in phytochemical processing. Forests are the main source of collection of many medicinal plants. However, the systematic cultivation of medicinal plants is also increasing. Systematic cultivation will reduce the pressure on forests and will also ensure repeatability in concentration. A preliminary knowledge of the plant parts to be collected is very useful. Plants/parts that are free from diseases and microbial infection should be collected. In compliance with good manufacturing practices (GMP), good agriculture collection practices (GACP) should be followed for the collection of plant materials.

2.2.2 Authentication of plant material

Collected plant material should be authenticated by a botanist or plant taxonomist to establish its botanical identity. Ideally, a voucher specimen should be deposited in a herbarium so that, in the future, reference could be made for the plant studied.

2.2.3 Cleaning of plant materials

Contamination can be avoided by cleaning the plant material. Generally, collected plant material is washed with water or treated with surfactants to remove dirt, grease and other sticky substances.

2.2.4 Drying and storage

Plant material is dried before extraction under controlled conditions to avoid chemical changes. High temperature is avoided, and the drying techniques commonly used are sun drying, air drying, solar drying and vacuum drying. Thoroughly dried plant material can be stored for a long time before extraction, either in a dry atmosphere or at a low temperature.

2.3 Extraction Parameters

The classical procedure for obtaining pure compounds from natural sources (plant, marine or animal) is extraction with a range of solvents, either with a single solvent or a combination of solvents. The mode of extraction depends on the substance that is being isolated. Extraction is followed by filtration with some inert material like Celite® (to remove foreign materials). The extract obtained is further sequentially fractionated with solvents of increasing polarity, providing distribution of phytochemicals in non-polar, semi-polar and polar solvents.

2.3.1   Selection of extraction method

The important distinguishing features among the different methods are time and the amount of solvent needed for an exhaustive extraction. The choice of the method of extraction of analytes influences the results.

2.3.2   Selection of solvent

The selection of an appropriate solvent or mixture is the first step. Non-polar solvents like petroleum ether are excellent solvents for non-polar compounds; however, their ability to extract polar compounds is often poor. Matching the polarity of the compound of interest and the extraction solvent is critical for optimum extraction. Due to the presence of a wide number of metabolites, it is not possible to extract them all with one single solvent. Toxicity, solubilization power, selectivity and chemical reactivity are the important factors to which attention must be paid while selecting a solvent. In ideal cases, solvent should be non-toxic and have a high solubilizing power. Solubilizing power and selectivity are important characteristics of a solvent. Solubilizing power is the power to dissolve a solute, whereas selectivity is the ability to dissolve specific compounds when the polarities of the various compounds present are not very different. Synder (1978) calculated a polarity index for various solvents and suggested the following equation to adjust the polarity of different mixtures:

$$P' = \phi P_a + \phi P_b,$$

where ϕ is the volume fraction and P_a and P_b are the polarity indices of solvent a and solvent b.

Green solvent is a better choice over chlorinated and aromatic solvents. Solvents with a zero dielectric constant like petroleum ether are not suitable for microwave-assisted extraction (MAE). Therefore, solvents for MAE must have a dielectric constant so that the absorption of microwaves is sufficient. The next step is the optimum sample solvent ratio. The characteristics of the some commonly used solvents are summarized in Table 2.1.

2.3.3   Extraction at room temperature and high temperature

The effect of temperature is important on extraction efficiency as temperature affects both the solubility of the compound and the diffusibility of the

Table 2.1. Characteristics of some common solvents used for extraction.

Solvent	Boiling point (°C)	Density (g ml⁻¹)	Dielectric constant
Acetone	56	0.79	20.7
Acetic acid	118	1.05	6.2
Carbon disulfide	46	1.26	2.6
Chloroform	61	1.49	4.8
Cyclohexane	81	0.78	2.0
Dichloromethane	40	1.33	8.9
Ethyl acetate	77	0.90	6.0
Formic acid	101	1.22	58.5
Hexane	69	0.66	1.9
Ethanol	78	0.79	24.6
Methanol	65	0.79	32.7
n-Butanol	118	0.81	17.5
n-Propanol	97	0.80	20.3
Toluene	111	0.87	2.4
Water	100	1.0	80.2

solvent through the sample matrix. As compared to room temperature, the capacity of the solvent to solubilize analytes increases at higher temperature. The solubility of hydrocarbons increases several hundred times when the temperature of the solvent is increased from 5°C to 15°C. The solubility of water in organic solvents also increases with increasing temperature; hence the possibility of analytes to be dissolved in the solvent is enhanced. Temperature and extraction time (number of extraction cycles) must be effectively optimized. An increase of about 30% extraction efficiency of total phenolics from black cohosh was observed when the temperature was increased from 40°C to 90°C. However, a further increase of temperature reduced the yield by 20%, possibly due to the unstable nature of phenolics (Mukhopadhyay *et al.*, 2006).

2.3.4 Particle size and sample–solvent ratio

Particle size and sample–solvent ratio also affect solvent use and extraction efficiency, although the effects are not well documented in the literature. Decrease of particle size increases the surface area, allowing better interaction between the sample and the solvent, and thereby better extractability. Large variations in sample to solvent ratio have been reported (Luthria *et al.*, 2006).

2.4 Extraction Technique

Plants are known to provide substances with new structural and novel bioactive properties. Phytochemicals isolated from plant extracts can be

used as an excellent source of phytotherapies. Growing interest in plant
secondary metabolites has created the need for economical, rapid and efficient extraction methods. The potential of extraction technologies for
the efficient extraction of natural products may assist in expanding phytochemical analysis, as well as characterization of the biologically activities of these compounds. Rapid advancement in separation and analysis
procedures has led to the discovery of new natural products. A variety of
solid–liquid extraction (SLE) methods have been reported for the purification of phytochemicals from plants, but limited progress has been made
in improving the extraction procedure. The traditional SLE methods include maceration, percolation and solvent extraction methods based
mostly on solvent type. These extraction techniques have a long extraction time coupled with low efficiency, are mostly manually operated and
require a large quantity of solvents. The application of energy sources
like microwave and ultrasound has developed faster extraction processes.
Unconventional extraction methods applying modern technologies include accelerated solvent extraction (ASE), microwave-assisted solvent
extraction (MASE), ultrasound-assisted solvent extraction (UASE) and
supercritical fluid extraction (SFE). Many natural products are thermally
unstable and degrade during thermal extraction; therefore, the optimization of operating parameters for a specific plant matrix (solvent polarity,
sample–solvent ratio, temperature, time and pH) is essentially required
for the highest extraction efficiency. Besides this, the optimization of extraction methods allows reduction of the usage and handling of solvents
and the waste generated in the extraction process. It also increases sample
throughput and reduces operation costs. Extraction techniques, both conventional and unconventional, may be summarized as follows:

- aqueous and organic solvent extraction;
- Soxhlet and accelerated solvent extraction;
- microwave-assisted solvent extraction;
- ultrasound-assisted solvent extraction;
- supercritical fluid extraction; and
- hydrodistillation.

2.4.1 Aqueous and organic solvent extraction by maceration and percolation

The physiochemical properties of the compound to be extracted determine the selection of the solvent. If the compound of interest is highly
soluble in water, hot or cold water is used for extraction. When the compound is highly soluble in a particular organic solvent, organic solvent
extraction is carried out. Using mild physical conditions, infusion is prepared from plant tissues with hot or cold water. However, decoction is
obtained by applying relatively vigorous physical conditions of extraction
with boiling water and longer extraction times. Extraction with organic

solvents is also carried out at room temperature. Finely powdered plant material is soaked with organic solvent at room temperature and this is repeated with the same solvent at least three to four times for exhaustive extraction.

2.4.2 Soxhlet and accelerated solvent extraction

Soxhlet extraction is a standard method most frequently used for the determination of oil content. This technique is also being used extensively for the extraction of phytochemicals from plants using solid matrix. In terms of process, here, extraction efficiency is influenced by factors such as the polarity of the solvent, temperature, extraction time and sample–solvent ratio. The temperature of extraction depends on the boiling point of the solvent used. A Soxhlet apparatus consists of a solvent reservoir, an extractor body, an electric heat source and a water-cooled reflux condenser. Finely powdered plant material is absorbed on Celite® and placed in a Whatman paper thimble. The thimble is placed in a Soxhlet apparatus (Fig. 2.1) and extracted sequentially with hexane, toluene, chloroform, ethyl acetate and *n*-butanol. Evaporation of the solvents under reduced pressure yields hexane, toluene, chloroform, ethyl acetate and *n*-butanol extracts. This process is less complicated compared to liquid–liquid extraction since water is not used for partitioning. Also, no emulsion formation takes place at any stage of the extraction; there is no need to dry

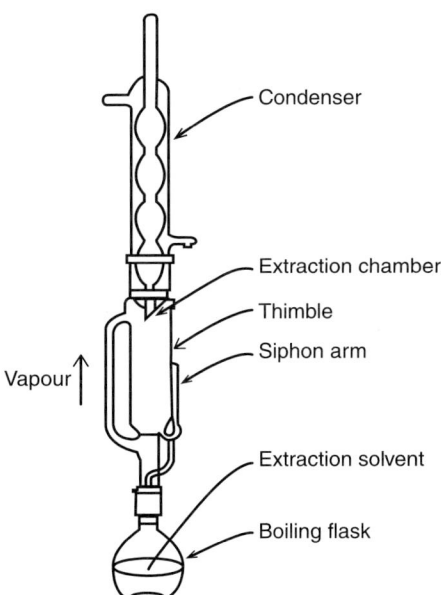

Fig. 2.1. Soxhlet extractor for extraction of phytochemicals.

the fractions before concentration, and in most cases yield is higher compared to a liquid–liquid partitioning process.

As compared with modern extraction techniques such as MASE and SFE, it is possible to extract a large number of samples using Soxhlet (Luque de Castro and García-Ayuso, 1998). Longer extraction time, large quantities of solvent and the possible degradation of thermolabile compounds because of continuous heating are the major drawbacks of the Soxhlet extraction technique.

Accelerated solvent extraction (ASE), also known as pressurized solvent extraction (PSE), is a new extraction technique similar in principle to Soxhlet extraction. It employs a combination of increased temperature and pressure. The use of elevated temperature and pressure with ASE allows the extraction to be completed in a short time with a small volume of solvent (Schafer, 1998). Increased temperature accelerates extraction kinetics as the diffusion rate is increased and decreases the viscosity of the solvent, thereby allowing better penetration of the matrix and weakening of the solute–matrix interaction. Elevated pressure controls the boiling point of the solvent and forces the solvent into matrix pores, facilitating better extraction (Matthaus and Bruhl, 2001; Kaufmann and Christen, 2002). Here, the solvents used are also those normally used for standard liquid extraction techniques like Soxhlet. Optimization of temperature, pressure, time and volume of solvent is important. There are a number of advantages in working with pressurized hot solvents, such as the solvent properties can be tuned by changing temperature, which in turn affects the dielectric constant, diffusion rate, viscosity and surface tension of the solvent. This results in a fast and efficient extraction. Water at elevated temperature (above its boiling point) is known to have a similar dielectric constant to an organic solvent such as methanol and acetonitrile, and it is likely that the property of water is comparable to that of an organic solvent. It has been reported that pressurized hot water can be used to extract moderately polar to polar compounds from plant materials. Superheated water (water under pressure and above 100°C but below its critical temperature (T_c = 374°C)) has been used for the extraction of herbal samples (Basile *et al.*, 1998; Ammann *et al.*, 1999; Clifford *et al.*, 1999). Although superheated water has advantages such as higher extraction ability, it is not suitable for thermally labile compounds. Further, oxygen should be purged carefully otherwise water at high temperature could damage the extraction vessel due to corrosive action.

The first instrument of ASE was commercialized in 1994 by the Dionex Corporation (Sunnyvale, California) (Reighard and Olesik, 1996; Richter *et al.*, 1996). With the recent development of an automated solvent dispenser, more than one solvent can be dispensed in the extraction vessels. Extraction time and volume of solvent consumption in ASE are much less than in the Soxhlet extraction.

ASE was used for the extraction of essential of thyme (*Thymus vulgaris*), phenolic compound from the bark of the Osage orange tree (*Maclura pomifera*), alkaloids from coca leaves and xanthones and taxnes (paclitaxel, baccatin III and 10-deacetylbaccatin III) from the bark of *Taxus cuspidata* (Benthin *et al.*, 1999; Da Costa *et al.*, 1999; Kawamura *et al.*, 1999; Brachet *et al.*, 2001). Paclitaxel, having poor solubility in water, exhibited about 50 times higher recovery in ASE than with cold water and about five times higher than with hot water (Kawamura *et al.*, 1999). Some of the applications of ASE in the isolation of phytochemicals are summarized briefly in Table 2.2.

2.4.3 Microwave-assisted solvent extraction (MASE)

Microwaves have been recognized as an outstanding source of energy to promote extraction. Microwave-assisted solvent extraction, developed in the late 1980s, is an alternative to conventional extraction techniques for natural products. In this method, extraction is carried out with a suitable solvent and microwave radiation. Yield is comparable and even better

Table 2.2. Applications of accelerated solvent extraction for the extraction of phytochemicals.

Phytochemicals	Plant/plant parts	Extraction conditions	Reference
Taxanes (paclitaxel, baccatin III and 10-deacetylbaccatin III)	*Taxus cuspidata* (Japanese yew) bark	Methanol–water (90:10); temperature: 150°C; pressure: 10.13 MPa	Kawamura *et al.*, 1999
Total anthocyanins and total phenolics	Dried red grape skin	Acidified water, methanol–acetone–water–hydrochloric acid (40:40:20:0.1); temperature: 50°C; pressure: 10.1 MPa	Ju and Howard, 2003
Antioxidants and polyphenols	Apple pomace	Ethanol (60%); temperature: 160–193°C/75–125°C; pressure: 10.3 MPa	Wijngaard and Brunton, 2009
Isoflavones	*Radix puerariae*	Ethanol (60%); temperature: 60–100°C; pressure: 1400 psi	Lee and Lin, 2007
Furanocoumarins	*Pastinaca sativa*	Petrol, methanol; temperature: 60–100°C; pressure: 60 bar	Waksmundzka-Hajnos *et al.*, 2004
Flavonoid glycosides and amino acids	*Scutellaria lanteriflora*	Water; temperature: 85–190°C; pressure: 10 MPa	Bergeron *et al.*, 2005
Cocaine and benzoylecgonine	*Erythroxylum coca*	Methanol; temperature: 80°C; pressure: 20 MPa	Brachet *et al.*, 2001
Silymarin	*Silybum marianum*	Water; temperature: 110–180°C	Bunnel *et al.*, 2010
Kavain	*Piper methysticum*	Acetone	Warburton *et al.*, 2007

than conventional methods, but in less time. Ganzler and co-workers were the first to report the use of microwave radiation for the extraction of natural products (Ganzler *et al.*, 1986a,b; Ganzler and Salgo, 1987). During MASE, the solvent is heated as a result of the absorption of microwave radiations by dipoles existing in the solvent or sample material. Microwave treatment leads to vaporization of water in the sample, destroying the cellular system and thereby enabling an exhaustive extraction. In contrast to conventional heating, no temperature gradient exists in the case of MASE; therefore, an even heating of the solution is ensured, thus facilitating the heating of the whole sample simultaneously. Disruption of the hydrogen bond due to the dipole rotation of the molecule is an added advantage in the case of extraction using microwaves. For MASE, the sample and/or solvent must have sufficient dielectric constant as microwave absorption ability depends on the dielectric constant. The larger the dielectric constant of the solvent, the more optimal the heating. Also, in some cases the matrix itself interacts with the microwaves as the surrounding solvent has a low dielectric constant (Jassie *et al.*, 1997). Solvents used for MASE cover a wide range of polarity; however, to improve extraction selectivity and the ability of the medium to interact with the microwave, a combination of more than one solvent can be used (Renoe, 1994). Water has the highest microwave absorption, whereas hexane is literally transparent to microwaves and thus does not heat up (Tatke and Jaiswal, 2011). Consequently, the moisture content in the sample matrix promotes better extraction as water superheats locally and promotes the release of analytes into the surrounding medium. Further control of the moisture of the matrix ensures reproducible extractions (Onuska and Terry, 1993; Budzinski *et al.*, 1996; Jassie *et al.*, 1997).

Microwave radiation is usually applied for short intervals, followed by an interval for cooling so that overheating is avoided. The temperature is also monitored externally by infrared sensor. Two types of instruments are available for MASE: (i) closed-vessel microwave extractor (CV-MAE); and (ii) focused open-vessel microwave extractor (FOV-MAE) (Fig. 2.2). Instrumentation is the main difference between CV-MAE and FOV-MAE. In closed-vessel extractors, the vessel containing the sample and solvent is closed, whereas in FOV-MAE it is open. In the case of CV-MAE, extraction is carried out under controlled pressure and temperature increase is achieved rapidly. Pressure depends on the volume and boiling point of the solvent as the solvents may be heated to about 100°C above their boiling point under atmospheric conditions in CV-MAE. The temperature can be fixed by adjusting the microwave power (100–1000 W), and power should be suitably chosen to avoid overheating, leading to degradation of analyte and overpressure. Both extraction efficiency and speed are reported to be enhanced in this procedure (Pare, 1990, 1991; Barnabas *et al.*, 1995; Young, 1995; Jassie *et al.*, 1997).

MASE in FOV-MAE is carried out at atmospheric pressure, and the maximum temperature depends on the boiling point of the solvent. The solvent is heated and refluxed through the sample. Here, the microwave is focused

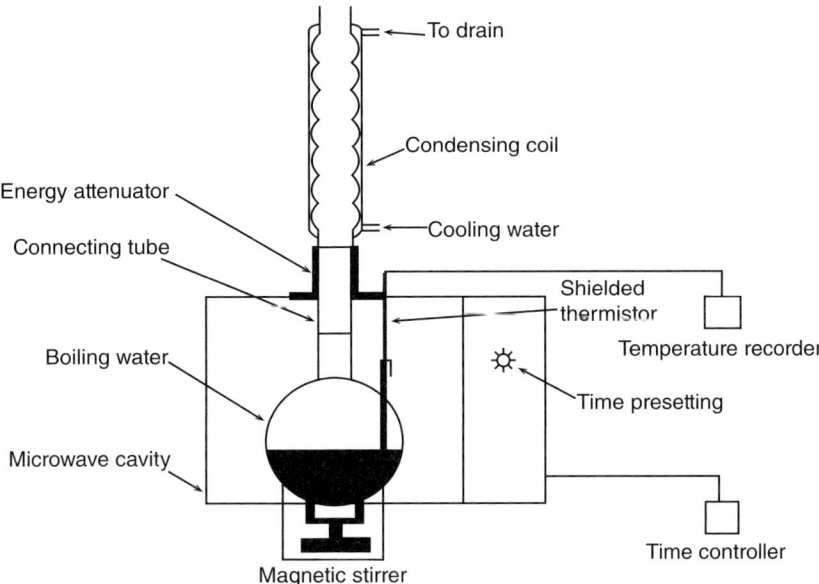

Fig. 2.2. Schematic diagram of an open-vessel microwave extractor. (From Mandal *et al.*, 2007.)

on the sample placed in the vessel; this generates homogeneous and effective heating. It has been reported that, as compared to the closed-vessel system, open-vessel MASE offers better safety and allows a larger sample to be extracted (Renoe, 1994; Letellier *et al.*, 1999; Kaufmann and Christen, 2002).

The MASE method could serve as a good alternative process for the preparation of high-value bioactive extracts, and the method has been utilized to extract a large number of bioactive phytochemicals from plant sources. Dai *et al.* (2001) reported the influence of various operating parameters, for example microwave power, solvent and irradiation time, on the recovery of azadirachtin-related limonoids (AZRL) from the seed kernel, seed shell, leaf and leaf stem of the neem tree. The extraction efficiency of MASE was compared with conventional extraction methods. MASE enhanced the extraction of AZRL from different parts possessing microstructure. Solvent also influenced the selectivity of microwave-assisted extraction. Cocaine and benzoyleconine were extracted from coca leaves using MASE (Brachet *et al.*, 2002). Quantitative recovery of cocaine from leaves was obtained in 30 s, and MASE-generated extract was similar to extract obtained from conventional extraction methods. Recoveries of lipophilic markers, mainly alkamide, from the roots of *Echinacera purpura* Monarch was evaluated by applying MASE (Hudaib *et al.*, 2003). Soxhlet and UASE reference methods were compared with MASE in terms of solvent and extraction time. Using methanol (70%) as the solvent, both MASE and UASE were found to be superior compared to Soxhlet. Both MASE and UASE were further evaluated by applying a different ratio of methanol–water (60–70%) as the solvent system. Recoveries were higher in MASE than in UASE, over 70–100% methanol range, while values were comparable at 60% methanol. The effects of the degree of grinding, solvent–material ratio and dielectric constant of solvent were optimized for the extraction of artemisnin from *Artemisia annua* (Hao *et al.*, 2002). Optimal conditions for MASE were: duration of irradiation, 12 min; diameter of raw materials, less than 0.125 mm; and solvent–sample ratio, more than 11.3. Under the optimized conditions of MASE, the yield (92.1%) was higher than with Soxhlet and normal stirring extraction (60%). Notably, supercritical fluid extraction using carbon dioxide had the lowest extraction yield of artemisnin. Four extraction methods including MASE were compared for recovery of the anticancer drug, camptothecin (CPT), from *Nothapodytes foetida*. The maximum percentage of extraction (2.67%) of CPT was obtained by MASE. The extraction time for MASE (3 min) was less than that for Soxhlet (120 min), UASE (30 min) and stirring extraction (30 min). Also, methanol (90%, v/v) yielded more extract than ethanol (90%, v/v). Shu *et al.* (2005) also reported that MASE (15 min, 150 W) yielded a better percentage of ginsenosides Rg_1 (0.28%, 70% methanol–water) and Rb_1 (1.31%, 30% water–ethanol) in comparison to 10 h solvent extraction (0.22% of Rg_1 and 0.87% of Rb_1 obtained in 70% water–ethanol). Similarly, in the case of the taxane class of natural products from the needles of *Taxus*, MASE reduced both extraction time and solvent consumption considerably while maintaining the qualitative and quantitative recovery of taxane relative to SLE methods. Casazza *et al.* (2010) compared SLE, UASE and MASE for the extraction and antioxidant power of phenolics from *Vitis vinifera* wastes. Although the highest content of total polyphenol, *o*-diphenols and flavonoids for seeds and skins were obtained using high pressure and temperature extraction, the highest antiradical power was determined in seed extracts obtained using MASE. For saponins extraction from chickpea (*Cicer arietinum*), MASE proved better compared with Soxhlet extraction. A butanol–water mixture exhibited selectivity towards saponin extraction. MASE contributed stability of heat-sensitive major saponins, that is, DDMP (2,3-dihydro-2,5-dihydroxy-6-methyl-4H-pyran-4-one)-conjugated saponins (Karem *et al.*, 2005). Liazid *et al.* (2007) investigated stability of 22 phenolic compounds of different families (benzoic acid, benzoic aldehydes, cinnamic acids, catechins, coumarins, stilbenes and flavonoids) under MASE at working temperatures between 50°C and 175°C. It was reported that all the compounds were stable up to 100°C, but at 125°C significant degradation of epicatechin, resveratol and myricetin was observed. Chemical structure and relationship studies revealed that compounds with a greater number of hydroxyl-type substitutions degraded more easily under extraction conditions.

For essential oil extraction from the leaves of *Rosamirunus officianals*, Bousbia *et al.* (2009) reported the microwave hydrodiffusion and gravity (MGH) technique and compared its effectiveness with hydrodistillation.

No solvents or water were used in MGH, and this method yielded essential oil containing a high amount of oxygenated compounds, thereby increasing the antimicrobial and antioxidant activities of the essential oil. Tigrine-Kordjani *et al.* (2006) also reported microwave-assisted, solvent-free distillation of essential oil from different aromatic plants. The direct interaction of microwaves with stem produced from the water present in fresh plant materials favours the release of essential oils trapped inside the cells of plant tissues. Dai *et al.* (2010) also reported the enhanced yield of three mint compounds, menthone, menthol and menthofuran, from peppermint leaves.

Ionic liquids, considered as 'green solvents', are also being used in microwave-assisted extraction. Du *et al.* (2009) used ionic liquids (ILs) for the extraction of phenolic compounds from *Psidium guajva* leaves and *Smilax china* tubers, and reported that ILs could replace efficiently the conventional reflux method with methanol. Some of the applications of MASE for the extraction of phytochemicals are summarized in Table 2.3, and comparisons of MASE with some other extraction methods are summarized in Table 2.4.

2.4.4 Ultrasound-assisted solvent extraction (UASE)

As with microwaves, ultrasound (frequency range 18–40 kHz) has been recognized as an excellent energy source for promoting extraction. UASE is much faster than conventional extraction methods due to the high contact surface area between solid and liquid phase. High-frequency sound

Table 2.3. Application of MASE for the extraction of phytochemicals.

Phytochemical	Plant	Solvent	Reference
Azadirachtin-related limonoids	*Azadirachta indica*	Methanol–water	Dai *et al.*, 2001
Cocaine and benzoylecgonine	Coca leaves	Methanol	Brachet *et al.*, 2002
Alkamides	*Echinacea purpurea* (L.) Monarch	Methanol–water	Hudaib *et al.*, 2003
Ginsenosides	*Panax ginseng*	Methanol–water, ethanol–water	Shu *et al.*, 2005
Camptothecin, 9-methoxycamptothecin	*Nothapodytes foetida*	Methanol (90%), ethanol (90%)	Fulzele and Satdive, 2005
Taxanes	*Taxus* spp.		Mattina *et al.*, 1997
Saponins	*Cicer arietinum*	*n*-Butanol and water mixture	Karem *et al.*, 2005
Essential oil	*Rosmarinus officinalis*	No solvent	Bousbia *et al.*, 2009
Glycyrrhizic acid	Licorice root	Ethanol (50–60%, v/v)	Pan *et al.*, 2000

Table 2.4. Comparison of Soxhlet extraction, MASE, ASE and SFE.

Factor	Soxhlet	MASE	ASE	SFE
1. Investment	Small	Medium	Large	Large
2. Process time	Long (up to 48 h)	Short (<30 min)	Short (<30 min)	Short (<60 min)
3. Solvent consumption	High (200–500 ml)	Low (40 ml)	Medium (<100 ml)	Minimal (<5 ml)
4. Method development	Simple	Simple	Simple	Labour-intensive
5. Sample treatment	Required	Required	Required	Not required

energy releases phytochemicals from plant materials through cavitation. Cavitations, or the formation and collapse of microscopic bubbles, release tremendous energy as heat, pressure and mechanical shear, thereby enhancing mass transfer and facilitating solvent access to the cell (Chemat *et al.*, 2004; Novak *et al.*, 2008). Ultrasound energy has been shown to improve extraction from vegetal tissues through the action of accelerating the rehydration or swelling of the plant cell that is accompanied by the fragmentation of the tissue matrix through mass transfer and penetration of the solvent into the cell, thereby promoting the absorption of the cell contents in the solvent (Vinatoru *et al.*, 1997; Toma *et al.*, 2001). The amount of ultrasonic cavitations in the solvent mixture is affected by surface tension, viscosity and vapour pressure (Chen *et al.*, 2007). Low vapour pressure solvents produce few cavitations, whereas high vapour pressure solvents produce more cavitations, but they collapse with less intensity and are therefore not very effective in extraction. Cavitations occur more easily in low-viscosity solvents, and also the ultrasonic intensity applied could exceed the molecular forces of the solvent more easily. As low-viscosity liquids have lower density and high diffusivity, they can diffuse easily in the pores of plant materials. Surface tension also influences cavitational effects. Low surface tension solvents require low energy to produce cavitations (Mason *et al.*, 1996; Gutiérrez *et al.*, 2008). Ultrasound-assisted extraction (UAE) is fast compared with traditional extraction methods like Soxhlet and cold percolation methods, and has the potential to increase analytical throughput while keeping solvent volumes and sample masses low.

In this method, finely powdered plant material mixed with solvent is sonicated in an ultrasonicator for about 10–60 min at a particular temperature (Fig. 2.3). If the compound is thermally unstable, extraction is carried out at low temperature to avoid thermal degradation or damage. Non-polar, polar and a combination of non-polar and polar solvents can be used in UASE. A combination of non-polar and polar solvents with ultrasonic energy has been reported to act like an emulsification–extraction, resulting in rapid and efficient extraction of total lipids from solid matrices (Perez-Serradila *et al.*, 2007). Ultrasonic power, volume and polarity of solvent, solvent and sample ratio, extraction time and pH are the important parameters influencing UASE.

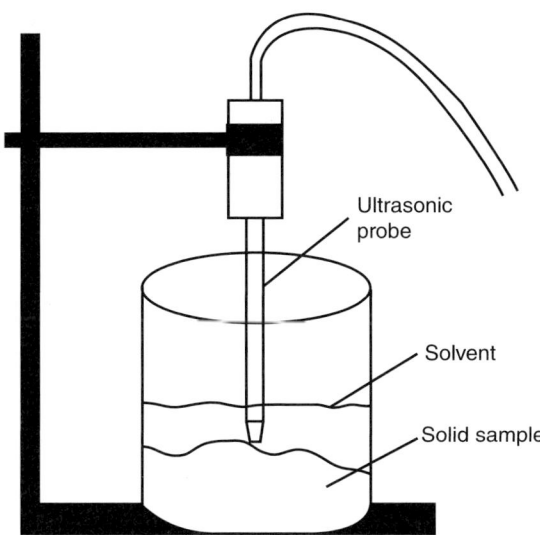

Fig. 2.3. Schematic diagram of an ultrasonic extractor. (From Kou and Mitra, 2003.)

The effects of ultrasonics have been studied for more than 100 plant species, and UASE has been reported for the extraction of phenols, ginsenosides, anthraquinones and polycylic hydrocarbons (Wu *et al.*, 2001; Chriestensen *et al.*, 2005; Hemwimol *et al.*, 2006; Richter *et al.*, 2006; Ahh *et al.*, 2007). Quan *et al.* (2009) reported that UASE was highly efficient in the extraction of ferulic acid from *Angelica sinensis*. Chen *et al.* (2007) reported that UAE was more efficient and rapid for the extraction of anthocyanins from red raspberry in comparison to conventional solvent extraction, possibly due to strong disruption of the fruit tissue structure under ultrasonic acoustic cavitation. The findings were further corroborated by scanning electron microscopic (SEM) studies. However, no changes in the composition of anthocyanins, as confirmed by high-performance liquid chromatography (HPLC), were observed. Conventional extraction and UASE were compared for the extraction of steroids and triterpenoids from three *Chresta* species, namely *C. exsucca*, *C. scapigera* and *C. sphaerocephala* (Schinor *et al.*, 2004). Total extraction time was reduced significantly in the case of UASE, and also this method was reported to be more effective for the extraction of steroids and most of the triterpenes. Li *et al.* (2005) reported that UASE was highly efficient for the extraction of chlorogenic acid from *Eucommia ulmoides* and also from other Chinese medicines as compared to classical methods. The influence of four variables was investigated with regard to the extraction efficiency of chlorogenic acid from fresh leaves and fresh and dried barks of *E. ulmoides*. Aqueous methanol (70%) in a solvent–sample ratio equal to 20:1 (v/w) with an extraction time of 90 min (3 × 30 min) were the optimum extraction conditions. The extraction efficiency of ethanol was improved for the isolation of carnosic acid from *Rosmarinus officinalis*. Ethanol, which is a poor solvent for the extraction of carnosic acid under conventional conditions, had a similar level of extraction to that of ethyl acetate and butanone in MASE. The yield of carnosic acid was also improved and the extraction time was shortened at the same temperature (Albu *et al.*, 2004). Stavarache *et al.* (2005) reported that the base-catalysed transesterification of vegetable oil with short chain alcohols with ultrasound (28 kHz and 40 kHz) was much shorter than with mechanical stirring. The quantity of catalyst required for transesterification was also reduced by two to three times. Forty kilohertz ultrasound was found much more effective in shortening the reaction time, but the yield was better at 28 kHz. However, higher frequencies were not useful for the transesterification of fatty acids. Extraction temperature is a factor that must be taken into account. This can be illustrated with the extraction of saponins from ginseng roots. UASE was found to be more efficient and three times faster than conventional Soxhlet extraction (Wu *et al.*, 2001). Together with total saponin content, the yield of ginsenosides Rb_1, Rb_2, Rc, Rd and Rf were determined individually after using both an ultrasonic bath (38.5 kHz, 810 W) and an ultrasonic probe (3 mm diameter tip, 20 kHz, 600 W) during extraction. Formation of ginsenoside Rg_3 and Rh_2 was reported because of thermal extraction process from the more abundant ginsenosides, Rb_1 and Rc (Popovich and Kitts, 2004).

The application of ultrasound irradiation facilitates low temperature rupturing of plant cell membranes, thereby liberating molecules from cellular structures, but extract cannot be separated completely from the solvent at the end. In general, UASE is carried out at a lower temperature than other forms of extraction, and this helps to avoid the degradation of thermally unstable ingredients in plant materials. Some applications of UASE for the extraction of phytochemicals are summarized in Table 2.5.

2.4.5 Supercritical fluid extraction (SFE)

Supercritical fluids as solvents have advantages, such as excellent mass transfer and control of solubility by temperature and pressure. Supercritical fluid extraction (SFE) has been used for the extraction of organic compounds from natural products, and industrial-scale plants using SFE have been employed for the decaffeination of coffee and tea and also the extraction of bitter principles from hops (Laws *et al.*, 1980; Zosel, 1981; Lack and Seidlitz, 1993). Unlike most organic solvents, carbon dioxide is not environmentally harmful and the advantages of supercritical carbon dioxide may be summarized as:

1. Low viscosity and high diffusion coefficient makes extraction and separation of short duration.
2. By regulating temperature and pressure, selective separation is feasible.
3. Oxidation and thermal degradation can be prevented as extraction is possible under mild conditions without oxygen.
4. As carbon dioxide is a gas at room temperature, solvent elimination and condensation is easier.

Table 2.5. Extraction of phytochemicals using UASE.

Compound	Plant/parts	Solvent and extraction time	Yield	Reference
Ferulic acid	*Angelica sinensis*	Ethanol (8 ml/g); 30 min	Yield and content were higher than percolation and supercritical fluid extraction	Quan *et al.*, 2009
Curcuminoids	*Curcuma longa* L.	Ethanol–water (1:1); 15 min	Yield was three times higher than conventional extraction	Rouhani *et al.*, 2009
Anthraquinones	*Morinda citrifolia*	Acetone, acetonitrile, methanol, ethanol	Higher recovery than Soxhlet and maceration	Hemwimol *et al.*, 2006
Alkaloids	*Hyoscyamus muticus*, *Datura stramonium* and *Ruta graveolens*	Surfactants; 2.5 h	Higher yield than Soxhlet	Djilani *et al.*, 2006
Anthocyanins	*Rubus idaeus*	Solvent (4 ml/g); 200 s	More efficient and rapid than conventional extraction	Chen *et al.*, 2007
Steroids and triterpenes	*Chresta* species		More efficient for extraction of triterpenes	Schinor *et al.*, 2004
Flavonoids and sesquiterpene lactones	*Lychnophora ericoides* Mart	Mixture of immiscible solvents		Sargenti and Vichnewski, 2000
Chlorogenic acid	*Eucommia ulmoides* Oliv.	Aqueous methanol (70%); 90 min (3 × 30)	Highly efficient for extraction of chlorogenic acid as compared to classical extraction methods	Li *et al.*, 2005
Carnosic acid	*Rosmarinus officinalis*	Butanone, ethyl acetate, ethanol	Extraction efficiency improved	Albu *et al.*, 2004

5. Safe operation is possible because carbon dioxide is not flammable.
6. Low running costs due to the low price of high-purity carbon dioxide.

As compared to carbon dioxide, supercritical water has a much higher critical temperature and pressure (347.2°C and 220.6 bars) but its effect as a solvent is greater. It can be used for a variety of reactions. Supercritical carbon dioxide (SC-CO$_2$) has virtually no surface tension; therefore, better penetration into plant matrix is possible as compared to liquid solvents. But, being non-polar in nature, it does not have a permanent dipole moment; its capacity to dissolve polar and high molecular weight compound is limited. The addition of polar solvent as a modifier can increase its polarity (Wright *et al.*, 1987; del Valle *et al.*, 2005; Ling *et al.*, 2007).

The efficiency of SFE depends on a number of factors. Sample preparation, extraction parameters and collection mode have been described as the experimental variables for SFE (Luque de Castro *et al.*, 1994; Lehotay, 1997; Cheah *et al.*, 2006). Although water is only about 0.3% soluble in SC-CO_2 (Lehotay, 1997), it could enhance the interaction of the analyte–modifier because water can open pores and swell the matrix, thereby allowing the fluid better access to the analyte and bringing the analyte out of the matrix. The presence of moisture in the sample can affect the SFE depending on the nature of the solute and the matrix properties. At higher pressure, water becomes more soluble in the supercritical fluid and thereby there is preferential partitioning of the polar compounds into the aqueous phase. This affects reproducibility and yield, as reported in the case of the extraction of digoxin from *Digitalis lanata* leaves (Moore and Taylor, 1996). The level of moisture in the sample can be controlled by heating the sample or by adding drying agent (Burford *et al.*, 1993). Increasing pressure at a constant temperature increases the density and solvating power (Taylor, 1996), and thereby there is better extraction of both desired and undesired constituents. Swelling of the sample matrix after exposure to pressurized carbon dioxide may also hinder the disruption of the analyte–matrix complex and adversely affect extraction efficiency (Benner, 1998). Decreased yield was reported in the case of basil, lovage, stevia and carrots, due to alteration in the plant matrix caused by high-pressure carbon dioxide (Smith and Burford, 1992a,b; Barth *et al.*, 1995; Yoda *et al.*, 2003; Menaker *et al.*, 2004). The effects of temperature on the solvating power of supercritical fluids depend on analyte volatility and the density of the fluid (Taylor, 1996). Increasing the temperature at low pressure reduces the solvating power of supercritical fluid in the extraction of artemisinin from *A. annua* and ginkogolides from *Ginko biloba* (Chiu *et al.*, 2002; Xing *et al.*, 2003). However, the opposite was observed at high pressure; here, increased temperature at increased pressure enhanced solvating power (Luque de Castro *et al.*, 1994; Hawthorne and King, 1999; Rompp *et al.*, 2004). These observations established that the effect of density was more predominant at low pressure, while the effect of volatility was greater at high pressure. Although at higher temperature desorption, diffusion and dissolution of the analyte is favoured, at the same time, high extraction temperature may not produce a better yield, possibly due to degradation (Liu *et al.*, 1995; Pourmortazavi *et al.*, 2003; Rompp *et al.*, 2004). Careful modulation of pressure and temperature allows fractional separation through sequential separating vessels (Yamini *et al.*, 2002). Vindoline was selectively extracted from 100 indole alkaloids present in *Catharanthus roseus* leaves and carotenoids from chlorophyll from the microalgae, *Nannochloropsis gaditana* (Song *et al.*, 1992; Macias-Sanchez *et al.*, 2005).

A polar modifier is usually added to supercritical fluid to improve the solubility of more polar analyte or competitively displace the analyte from the matrix active sites (Hawthorne and King, 1999; Jeong and Chesney, 1999). The effect is dependent on the concentration of the modifier in the supercritical phase, which is determined by the phase behaviour of the mixture under operating conditions. Modifiers could also interact directly with the matrix–analyte complex to lower the energy barrier for desorption (Alexandrou *et al.*, 1992). The addition of modifiers increased the yield of the desired component by up to three times (Choi *et al.*, 1998; Wang *et al.*, 2001a). Methanol, having excellent hydrogen-bonding and proton-accepting properties, is the most commonly used modifier for extracting polar analytes. Ethanol, comparatively less polar than methanol but less toxic, has also been used for several extractions. In some cases, water alone has been used as a modifier, such as in the extraction of caffeine from coffee beans and tea leaves. The combination of modifiers can further improve the extraction efficiency of analytes. A combination of methanol and water was more effective than methanol in the extraction of polar flavonoid from *Scuttellariae radix* (Lin *et al.*, 1999) and cocaine from coca leaves (Brachet *et al.*, 2002). A method for predicting the solubility of solids in modified SC-CO$_2$ has been developed recently and employed with moderate success in the selection of a suitable modifier for naproxen as a solubility enhancer for SC-CO$_2$ (Abaroudi *et al.*, 2002).

The addition of a large amount of modifier will change the critical parameters of the mixture (Pourmortazavi and Hajimirsadeghi, 2007). The addition of ethanol increased the bulk density of SC-CO$_2$ due to the higher density of the co-solvent and clustering of SC-CO$_2$ molecules around the co-solvent (Guclu-Ustundag and Temelli, 2005). The solubility of β-carotene, both in pure SC-CO$_2$ and ethanol-modified SC-CO$_2$, was measured at 40°C, 50°C and 60°C (Sovová *et al.*, 2001). It was found that ethanol as a modifier increased the solubility by one order of magnitude and the increase in solubility was proportional to the square root of the modifier concentration.

SFE may be carried out in three modes, namely static, dynamic and a combination of static and dynamic modes. A supercritical carbon dioxide fluid extractor system is shown in Fig. 2.4. In static extraction, a fixed amount of supercritical fluids interacts with sample matrix for a specified time. This mode is generally used if the extraction process is desorption or kinetics controlled and a low concentration of analyte is bound strongly to the analyte (Bjorklund *et al.*, 1998). The dynamic mode is used when the extraction is solution/elution controlled. Here, fresh extractant is passing through the sample continuously. Increasing the flow would improve extraction efficiency (Ashraf-Khorassani *et al.*, 1997). In some cases, sequential static and dynamic extractions produce better yield (Modey *et al.*, 1996). The static phase facilitates displacement of the analyte from the matrix, followed by a dynamic sweep of the cell (Luque de Castro *et al.*, 1994; Taylor, 1996).

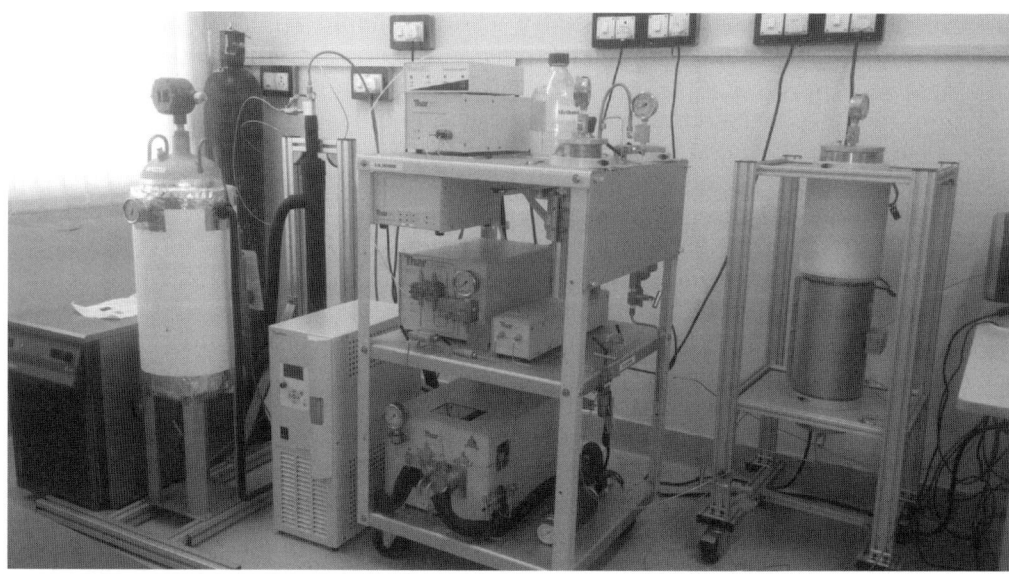

Fig. 2.4. Supercritical extraction facility at ICAR-Directorate of Medicinal and Aromatic Plants Research, Anand, India.

Three forms of collecting extracts from SFE extracts have been reported: (i) collection using a solid matrix; (ii) liquid collection using a small volume of solvent; and (iii) sublimation of carbon dioxide directly into a vessel. In solid collection, CO_2 and the extract are depressurized before collection, thereby enabling the adsorption of analytes on collection adsorbant or packing material. In liquid collection, CO_2 and analyte are depressurized and bubble directly through the outlet restrictor into a small amount of collection solvent. This method is useful when analytes are sensitive to oxidative degradation as antioxidant can be added to the solvent to protect oxidation (Turner *et al.*, 2002). CO_2 may also be depressurized directly into an empty flask where the analyte is deposited (Wright *et al.*, 1987).

The efficiency of the collection method is also controlled by the collection temperature. Sometimes, cooling of the collection vessel to subzero temperature is required to prevent the loss of volatile compounds (Smith and Burford, 1992a,b). In the case of high concentration of modifier, the temperature is elevated above the boiling point of the modifier to prevent condensation into the collection vessel. Finally, the ultimate choice of collection depends on the sample matrix, analyte type and extraction parameters.

SC-CO_2 has been used for the extraction and isolation of high-value phytochemicals from natural products during the last three decades (Martinelli *et al.*, 1991; del Valle and Aguilera, 1999; Hartono *et al.*, 2001). This technique was found to be suitable for the selective separation of desired constituents and has proved effective in the separation of essential oils, with more satisfactory composition (lower monoterpenes) than obtained from conventional hydrodistillation and decaffeination, thereby promoting commercial uses for the removal of caffeine from coffee and black tea (Ozer *et al.*, 1996; Diaz-Maroto *et al.*, 2002). Cholesterol is more soluble in SC-CO$_2$, and even more so in supercritical ethane; therefore, this technique has great potential for removing cholesterol from food products (Greenwald, 1991). Supercritical fluids also serve as a reaction media because of their capacity to homogenize the reaction mixture, high diffusivity and controlled phase separation and distribution of products (Phelps *et al.*, 1996). Applications of supercritical carbon dioxide in the isolation of phytochemicals have been summarized in Table 2.6.

2.4.6 Hydrodistillation

Essential oil can be isolated by hydrodistillation, steam distillation, as well as organic solvent extraction. Essential oil secreted by glandular trichomes is concentrated mainly in leaves, and hydrocarbons, esters, terpenes, phenols, carbonyl compounds and alcohols are their major constituents. Oxygenated compounds such as aldehydes, ketones, lactone, esters and alcohols are the source of fragrance and are more stable than unsaturated constituents like monoterpenes and sesquiterpenes because of the oxidizing tendency of the latter in the presence of air and light. Monoterpenes are well known to be vulnerable to chemical changes, even steam distillation (Presti *et al.*, 2005). Physical properties such as boiling point, thermal stability and vapour pressure–temperature relationship provide important information for optimizing yield efficiency.

For the isolation of essential oils using hydrodistillation, aromatic plant material is transferred to a distillation assembly and a sufficient quantity of water is added (Fig. 2.5). Temperature is adjusted to boil. Alternatively, steam may pass directly into the distillation assembly. Hot water or steam make the essential oil free of oil glands. A mixture of water and oil in the form of vapour is condensed by the circulation of water in the condenser. Distillate flows to the separator from the condenser, and there oil is separated automatically from distillate water. In spite of its simplicity and no involvement of solvent, steam distillation is a time- and energy-consuming technique. Also, high temperature and high moisture during the process may cause modification and degradation of the overall quality (Spiro and Chen, 1995).

2.4.7 Use of surfactants and detergents for extraction

The use of detergents, alcohols and dimethyl sulfoxide (DMSO) to extract compounds of interest without killing the tissue of living plants and seeds has been reported (Cseke *et al.*, 2006). The method relies on

Table 2.6. Applications of supercritical carbon dioxide in the extraction of phytochemicals. (From Cheah *et al.*, 2006.)

Phytochemical	Plant	Extraction conditions	Reference
Diosgenin (steroidal saponin)	*Dioscorea nipponica*	P = 3100 psi; 40 min	Liu *et al.*, 1995
Triterpenoids	*Smilax china*	P = 35 MPa; T = 65°C	Shu *et al.*, 2004
Lignans	*Schisandra chinensis*	P = 20–27 MPa; T = 40–65°C	Bartlova *et al.*, 2002
Podophyllotoxin	*Dysosma pleiantha*	P = 13.6–34 MPa; T = 40–80°C; modifier = methanol	Choi *et al.*, 1998
Artemisnin and artemisnic acid	*Artemisia annua*	P = 15 MPa; T = 50°C; modifier = methanol	Kohler *et al.*, 1997
Ginkgolides	*Ginko biloba*	P = 24.2–31.2 MPa; T = 60–120°C	Chiu *et al.*, 2002
Ginsenosides	*Panax ginseng*	P = 31.22 MPa; T = 60°C; modifier = methanol	Wang *et al.*, 2001b
Hyperforin	*Hypericum perforatum*	P = 120 bar; T = 40°C	Rompp *et al.*, 2004
Tagitinin C	*Tithonia diversifolia*	P = 35 MPa; T = 68°C	Ziemons *et al.*, 2005
Alkylamides	*Echinacea angustifolia*	P = 34–55 MPa; T = 45–60°C; modifier = methanol	Sun *et al.*, 2002
β-amyrin, β-sitosterol	*Taraxacum officinale*	P = 450 bar; T = 65°C; modifier = ethanol	Simandi *et al.*, 2002
Flavonoids, sesquiterpenes	*Chamomile* flowers	P = 90–200 atm; T = 40–45°C	Scalia *et al.*, 1999
Coumarins	*Mikania glomerata*	P = 10.1 MPa; T = 70°C; modifier = ethanol	Vilegas *et al.*, 1997
Pyrethrins	*Pyrethrum*	P = 1200 psi; T = 40°C	Pan *et al.*, 1995
Vindoline	*Catharanthus roseus*	P = 300 bar; T = 35°C	Song *et al.*, 1992
Paclitaxel	*Taxus cuspidata*	P = 300 bar; T = 35°C; modifier = dichloromethane	Ahsraf-Khorassani *et al.*, 1992
Glycoside	*Stevia rebaudiana* Bertoni	P = 200 bar; T = 30°C; modifier = ethanol	Yoda *et al.*, 2003
Digexin	*Digitalis lanata*	P = 380 bar; T = 100°C; modifier = methanol	Moore and Taylor, 1996
Nimbin	Neem seeds	P = 23 MPa; T = 35°C; modifier = methanol	Tonthubthimthong *et al.*, 2004
Resveratrol	*Vitis vinifera*	P = 150 bar; T = 40°C; modifier = ethanol	Pascual-Marti *et al.*, 2001
Squalene	*Terminalia catappa*	P = 300 psi; T = 40°C	Ko *et al.*, 2002
Lycopene	Tomato	P = 450 psi; T = 65–70°C	Vasapollo *et al.*, 2004

Note: 1 bar ≡ 100,000 Pa = 14.5038 psi.

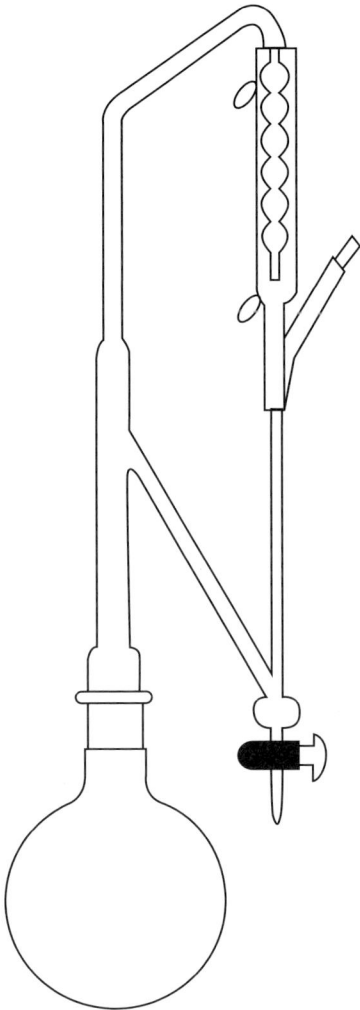

Fig. 2.5. Hydrodistillation assembly (Clevenger apparatus).

the partition coefficients of the chemicals used, the polarity of the molecules to be extracted and the ability of the solvents to penetrate the tissue. Solvents such as alcohol (methanol, ethanol, long chain alcohols) and detergents (XAD-4, vinyl benzene) were used. Viability tests must be run with the plants (Komolpis *et al.*, 1998; Wang *et al.*, 2001a). To extract two isoflavonoids, daidzein and genistein, from soybean seeds, Wang *et al.* (2001a) tested aqueous methanol and reported the effect of concentration of the solvent on viability in terms of percentage germination of the treated seeds. It was reported that the release of metabolites increased with increased percentage of methanol (0–30% aqueous methanol), possibly due to the increase in the solubility of the stored daidzen and genistein. Viability of the seeds dropped with increase in methanol concentration beyond 20%. Seed germination was above 80% (0–20% aqueous methanol, 24 h); however, no germination was observed at 30% aqueous methanol concentration.

Non-ionic surfactants have been used for extraction. Above a critical temperature, non-ionic surfactants show unique cloud point behaviour, that is, the ability to separate into two distinct phases, a surfactant-rich phase and a bulk aqueous phase. Fang *et al.* (2000) applied this principle in combination with UAE for the extraction of ginsenosides. Pressurized liquid extraction in combination with cloud point technology has been reported by Choi *et al.* (2003). As compared to water or methanol, the presence of the non-ionic surfactant, Triton X-100, at a concentration above its critical micelle concentration (CMC) increased the extraction of ginsenoside Rg_1, Re, Rb_1, Rc and Rd from ginseng roots under pressurized liquid extraction when compared with UAE (Choi *et al.*, 2003).

2.4.8   Use of ionic liquid for extraction

Ionic liquids (ILs) have recently attracted research interest for a variety of applications. ILs are composed of bulky organic cations and inorganic or organic anions present in liquid form at low temperature (<100°C). IL possesses negligible vapour pressure, good thermal stability, wide liquid range, tunable viscosity and miscibility with water as well as organic solvents. IL has been used in extractions and reactions. ILs are called green solvents and have been investigated extensively as solvents or co-solvents for extraction (Zhao *et al.*, 2005).

Ionic liquid-based, ultrasound-assisted extraction (ILUAE) has been applied successfully to the extraction of three alkaloids, vindoline, catharanthine and vinblastine, from *Catharanthus roseus*. Twelve ILs, with different cations and anions, were investigated in this work and [Amim] Br was selected as the optimal solvent. Ultrasound power and time, and the number of extraction cycles, were optimized. ILUAE offered short extraction times (from 0.5 h to 4 h) and remarkable efficiency (Yang *et al.*, 2011).

The ionic liquids-based, microwave-assisted extraction (ILs-MAE) technique was developed for the effective extraction of podophyllotoxin from three Chinese medicinal plants. 1-Butyl-3-methylimidazolium tetrafluoroborate ([bmim][BF4]), 1-decyl-3-methylimidazolium tetrafluoroborate ([demim][BF4]) and 1-allyl-3-methylimidazolium tetrafluoroborate ([amim][BF4]) were selected as the optimal surfactants for *Dysosma versipellis*, *Sinopodophyllum hexandrum* and *Diphylleia sinensis*, respectively. Compared with other extraction techniques, such as ILs-based maceration extraction, heat extraction and UAE, the ILs-MAE technique not only took a shorter time but also afforded a higher extraction rate of podophyllotoxin from the herbs.

2.4.9 Biorefinery

Plants are composed of a large number of different molecules like crude oil. Each component of the plant can be extracted and functionalized in order to produce non-food and food fractions and agroindustrial intermediate products. Carbohydrate, lignin, proteins and fats constitute about 95% of plants. Vitamins, dye, flavours or other small molecules are also considered in biorefinery because of their high value. Different specific biorefineries can be outlined based on sugars (starch and sucrose), lignocelluloses and lipids as the main source of carbon molecules (Octave and Thomas, 2009).

Similar to petroleum refineries, biorefineries process bioresources such as agriculture or forest biomass to produce energy and a wide range of precursor chemicals and bio-based materials. Biorefineries use a variety of separation methods, often to produce high-value co-products from the different types of feed streams. From wood and other lignocellulosic biomass, chemicals like acetic acid, liquid fuel such as bioethanol and biodegradable plastics such as polyhydroxyalkanoates can be produced.

2.5 Partitioning of Extract

Partitioning of the extract with solvents of varying polarity (Fig. 2.6) is followed by chromatographic techniques like open-column chromatography,

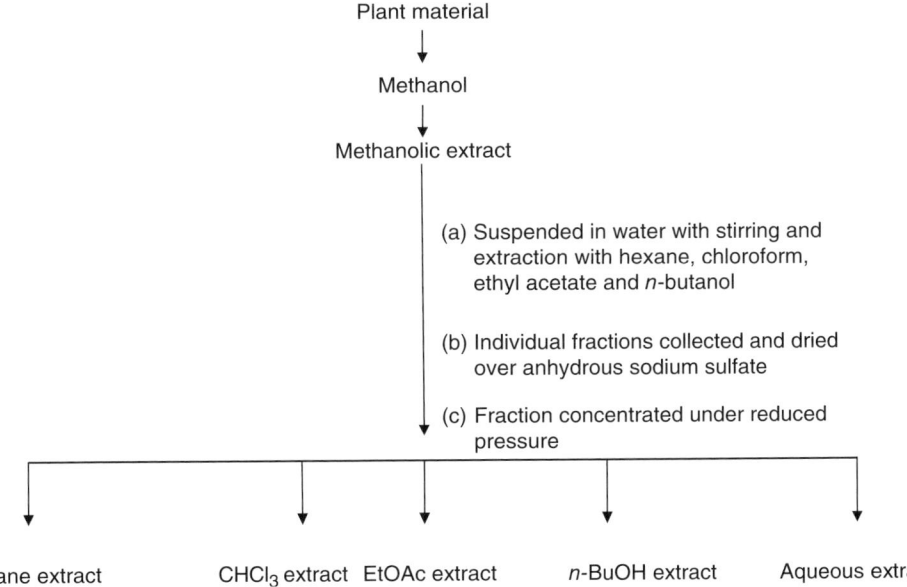

Fig. 2.6. Schematic diagram of portioning of methanol extract with solvents of varying polarity.

medium-pressure chromatography and preparative chromatography (thin-layer chromatography (TLC), HPLC) to complete phytochemical profiling.

2.6 Drying or Concentrating Extract

Solvent removal from the extract is commonly carried out using evaporation, vacuum concentration, lyophilization, etc. The objective of solvent removal is to preserve the solute. The removal of solvent can be accomplished effectively by boiling. Solvent can be boiled off either by applying heat or by lowering the atmospheric pressure. In both cases, the energy of molecular motion is greater than the intermolecular forces holding the molecules in the solution, and as a result solvent molecules move from liquid phase to gaseous phase. Similar to vacuum concentration, the process of lyophilization is carried out by lowering the temperature to the point where the solution freezes and the solvents are removed by sublimation. Lyophilization is very effective for concentrating and preserving biologically active extracts.

Spray drying is a well-established method of transforming a wide range of liquid products into powder form. Spray-dried powders are suitable for transport and storage. Spray drying is widely used in phytochemical processing to produce extract powders, mainly due to its relatively shorter process time and lower process economics as compared to other drying techniques such as freeze drying. The spray-dried extract is sold as the final product in bulk form (Athimulam *et al.*, 2006).

2.7 Upscaling of Process from Bench Scale to Pilot Plant

The standardization process ensures consistency in the active compound content of herbal extract or phytochemicals. The increase in the value of herbal products as processing and standardization increase is shown in Fig. 2.7 (Ismail, 2003). Optimized extraction parameters are used for

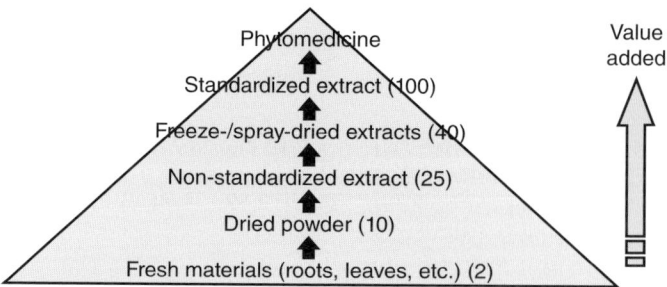

Fig. 2.7. Increasing the value of herbal products with processing and standardization. (From Ismail, 2003.)

the industrial-scale production of extracts. In the laboratory, extraction parameters are generally optimized in sample weights varying from 100 to 500 g of plant parts. The optimized parameters for maximum yield of extracts are utilized in pilot plant scale processing of 15 kg or higher weight of plant materials.

2.8 Conclusion

Sample collection and extraction are the important initial steps of metabolomics experiments. Validation of the extraction step is important as any non-extracted metabolites may affect the overall quality greatly. Therefore, it is vital to pay attention to the extraction protocol. An efficient extraction protocol should extract the largest number of metabolites and be non-destructive (Prasad and Ferenci, 2003; Mushtaq *et al.*, 2014). Along with proper solvent selection, both pre-extraction techniques and the treatment of the matrix during extraction play an important role in the release of metabolites. Further, to achieve good results, not only is the type of solvent used to be taken into account but also the physiochemical characteristics of the matrix, the effect of pH on the matrix, the contract time and compartmentalization of metabolites (Silas and Villas, 2006).

The use of unconventional extraction techniques has cut down the extraction time and increased the yield as well as the quality of the extract. MASE and UASE are exploited chiefly at the laboratory scale, and both techniques have been used for some industrial applications (Cravoto *et al.*, 2008). A statistical approach to the application of designs may further reduce the volume of solvent used as the number of experiments required for optimization is reduced (Alonso-Salces *et al.*, 2005). The selection of suitable modern extraction techniques and the optimization of extraction parameters will allow a reduction in solvent cost and waste generated during the extraction of plant materials. The use of green solvents such as water and carbon dioxide will reduce environmental pollution and health problems. The main impediments to the widespread application of modern extraction technologies remain the high investment costs and difficulties in the implementation of fully automated systems.

References

Abaroudi, K., Trabelsi, F. and Recasens, F. (2002) Screening of cosolvents for a supercritical fluid: a fully predictive approach. *AIChE Journal* 48, 551–559.

Ahh, Y.G., Shin, J.H., Kim, H.Y., Khim, J., Lee, M.K., *et al.* (2007) Application of solid phase extraction coupled with freezing lipid filtration clean-up for the determination of endocrine-disrupting phenols in fish. *Analytica Chimica Acta* 603, 67–75.

Albu, S., Joyce, E., Paniwnyk, L., Lorimer, J.P. and Mason, T.J. (2004) Potential for the use of ultrasound in the extraction of antioxidants from *Rosmarinus officinalis* for the food and pharmaceutical industry. *Ultrasonics Sonochemistry* 11, 261–265.

Alexandrou, N., Lawrence, M.J. and Pawliszyn, J. (1992) Cleanup of complex organic mixtures using supercritical fluids and selective adsorbents. *Analytical Chemistry* 64, 301–311.

Alonso-Salces, R.M., Barranco, A., Corta, E., Berrueta, L.A., Gallo, B., *et al.* (2005) A validated solid–liquid extraction method for the HPLC determination of polyphenols in apple tissues: comparison with pressurised liquid extraction. *Talanta* 65, 654–662.

Ammann, A., Hinz, D.C., Addleman, R.S., Wai, C.M. and Wenclawiak, B.W. (1999) Superheated water extraction, steam distillation and SFE of peppermint oil. *Fresenius Journal of Analytical Chemistry* 364, 650–653.

Anderson, R.A and Schmidt, W.F. (2000) Cinnamon extracts boosts insulin sensitivity. *Agricultural Research Magazine*, July, pp. 21.

Ashraf-Khorassani, M., Combs, M.T. and Tatlor, L.Y. (1997) Supercritical fluid extraction of metal ions and metal chelates from different environments. *Journal of Chromatography A* 777, 37–49.

Athimulam, A., Kumaresan, S., Foo, D.C.Y., Sarmidi, M.R. and Aziz, R.A. (2006) Modeling and optimization of *Eurycoma longigolia* water extract production. *Food and Bioproducts Processing* 84, 139–149.

Barnabas, I.J., Dean, J.R., Fowlis, I.A. and Owen, S.P. (1995) Extraction of polycyclic aromatic hydrocarbons from highly concentrated soils using microwave energy. *Analyst* 120, 1897–1904.

Barth, M.M., Zhou, C., Kute, K.M. and Rosenthal, G.A. (1995) Determination of optimum conditions for supercritical fluid extraction of carotenoids from carrot (*Daucus carota*) tissue. *Journal of Agricultural and Food Chemistry* 43, 2876–2878.

Bartlova, M., Opletal, L., Chobot, V. and Sovova, H. (2002) Liquid chromatographic analysis of supercritical carbon dioxide extracts of *Schizandra chinensis*. *Journal of Chromatography B* 770, 283–289.

Basile, A., Jimenez-Carmona, M.M. and Clifford, A.A. (1998) Extraction of rosemary by superheated water. *Journal of Agricultural and Food Chemistry* 46, 5205–5209.

Benner, B.A. Jr (1998) Summarizing the effectiveness of supercritical fluid extraction of polycylic aromatic hydrocarbons from natural matrix environmental samples. *Analytical Chemistry* 70, 4594–4601.

Benthin, B., Danz, H. and Hamburger, M. (1999) Pressurized liquid extraction of medicinal plants. *Journal Chromatography A* 837, 211–219.

Bergeron, C., Gafner, S., Clausen, E. and Carrier, D.J. (2005) Comparison of the chemical composition of extracts from *Scutellaria lateriflora* using accelerated solvent extraction and supercritical fluid extraction versus standard hot water or 70% ethanol extraction. *Journal of Agricultural and Food Chemistry* 53, 3076–3080.

Bjorklund, E., Jaremo, M., Mathiasson, L., Jonsson, J.A. and Karlsson, L. (1998) Illustration of important mechanisms controlling mass transfer in supercritical fluid extraction. *Analytica Chimica Acta* 368, 117–128.

Bousbia, N., Abert Vian, M., Ferhat, M.A., Petitcolas, E., Meklati, B.Y., *et al.* (2009) Comparison of two isolation methods for essential oil from rosemary leaves: hydrodistillation and microwave hydro diffusion and gravity. *Food Chemistry* 114, 355–362.

Brachet, A., Rudaz, S., Mateus, L., Chriestein, P. and Veuthey, J.L. (2001) Optimization of accelerated solvent extraction of cocaine and benzoylecgonine from coca leaves. *Journal of Separation Science* 24, 865–873.

Brachet, A., Chriesten, P. and Veuthey, J.L. (2002) Focused microwave assisted extraction of cocaine and benzoylecgonine from cocoa leaves. *Phytochemical Analysis* 13, 162–169.

Budzinski, H., Baumard, P., Papineau, A., Wise, S. and Garrigues, P. (1996) Focussed microwave assisted extraction of polycyclic aromatic compounds from reference materials, sediments and biological tissues. *Polycyclic Aromatic Compounds* 9, 225–232.

Bunnel, K.A., Wallace, S.N., Clausen, E.C., Penney, W.R. and Ca, J.C. (2010) Comparison of Silymarin extraction from *Silybum marianum* using a soxhlet apparatus, batch Parr, and countercurrent pressurized hot water reactors. *Transactions of the ASBAE* 53, 1935–1940.

Burford, M.D., Hawthorne, S.B. and Miller, D.J. (1993) Evaluation of drying agents for off-line supercritical fluid extraction. *Journal of Chromatography A* 657, 413–427.

Casazza, A.A., Aliakbarian, B., Mantegna, S., Cravotto, G. and Perego, P. (2010) Extraction of phenolics from *Vitis vinifera* wastes using non-conventional techniques. *Food Engineering* 100, 50–55.

Cheah, E.L.C., Chan, L.W. and Heng, P.W.S. (2006) Supercritical carbon dioxide and its application in the extraction of active principles from plant materials. *Asian Journal of Pharmaceutical Sciences* 1, 59–71.

Chemat, S., Lagha, A., Aitamar, H., Bartels, P.V. and Chemat, F. (2004) Comparison of conventional and ultrasound assisted extraction of carvone and limonene from caraway seeds. *Flavour and Fragance Journal* 19, 188–195.

Chen, F., Sun, Y., Zhao, G., Liao, X., Hu, X., *et al.* (2007) Optimization of ultrasound-assisted extraction of anthocyanins in red raspberries and identification of anthocyanins in extracts using high-performance liquid chromatography-mass spectrometry. *Ultrasonics Sonochemistry* 14, 767–778.

Chiu, K.L., Cheng, Y.C., Chen, J.H., Chang, C.J. and Wang, P.W. (2002) Supercritical fluids extraction of *Ginko ginkgolodes* and flavonoids. *Journal of Supercritical Fluids* 24, 77–87.

Choi, M.P.K., Chan, K.K.C., Leung, H.W. and Huie, C.W. (2003) Pressurized liquid extraction of active ingredients (ginsenosides) from medicinal plants using non ionic surfactant solutions. *Journal of Chromatography A* 983, 153–162.

Choi, Y.H., Kim, J.Y., Ryu, J.H., Yoo, K.P., Chang, Y.S., *et al.* (1998) Supercritical extraction of podophyllotoxin from *Dysoma pleiantha* roots. *Planta Medica* 64, 482–483.

Chriestensen, A., Ostman, C. and Westerholm, R. (2005) Ultrasound assisted extraction and on-line LC-GC-MS for determination of polycylic aromatic hydrocarbons (PAH) in urban dust and diesel particulate matter. *Analytical and Bioanalytical Chemistry* 381, 1206–1216.

Clifford, A.A., Basile, A., Al-Saidi, S.H.R. and Fresenius, J. (1999) A comparison of extraction of clove buds with super critical carbon dioxide and super heated water. *Fresenius's Journal of Analytical Chemistry* 364, 635–637.

Cravoto, G., Boffa, L., Mantegna, S., Perego, P., Avogadro, M., *et al.* (2008) Improved extraction of vegetable oils under high frequency ultrasound and/or microwaves. *Ultrasonics Sonochemistry* 15, 898–902.

Cseke, L.J., Setzer, N.J., Vagler, B., Kirakosyan, A. and Kaufman, P.B. (2006) Traditional, analytical and preparative separations of natural products. In: Cseke, L.J., Kirakosyan, A., Kaufman, P.B., Warber, S.L., Duke, J.A. and Brielmann, H.L. (eds) *Natural Products from Plants*, 2nd edn. CRC Press, Boca Raton, Florida, pp. 263–318.

Da Costa, C.T., Margolis, S.A., Benner, B.A.J. and Horton, D. (1999) Comparison of methods for extraction of flavanones and xanthones from the root bark of the *Osage orange* tree using liquid chromatography. *Journal of Chromatography A* 831, 167–178.

Dai, J., Yaylayan, V.A., Vijaya Raghvan, G.S., Jocelyn Pare, J.R., Liu, Z., *et al.* (2001) Influence of operating parameters on the use of the microwave-assisted process (MAP) for the extraction of azadirachtin-related limonoids from neem (*Azadirachta indica*) under atmospheric pressure conditions. *Journal of Agricultural and Food Chemistry* 49, 4584–4588.

Dai, J., Orsat, V., Raghavan, G.S. and Yaylayan, V. (2010) Investigation of various factors for the extraction of peppermint (*Mentha piperita* L.) leaves. *Journal of Food Engineering* 96, 540–543.

del Valle, J.M. and Aguilera, J.M. (1999) High pressure carbon dioxide extraction: fundamentals and applications in the food industry. *Food Science and Technology International* 5, 1–24.

del Valle, M.T., Rogalinski, T., Zentl, C. and Brunner, G. (2005) Extraction of boldo (*Peumus boldus* M.) leaves with supercritical CO_2 and hot pressurized water. *Food Research International* 38, 203–213.

Diaz-Maroto, M.C., Perez-Coello, M.S. and Cabezudo, M.D. (2002) Supercritical carbon dioxide extraction of volatiles from spices: comparison with simultaneous distillation extraction. *Journal of Chromatography A* 947, 23–29.

Djilani, A., Legsseir, B., Soulimani, R., Dicko, A. and Younos, C. (2006) New extraction technique for alkaloids. *Journal of Brazilian Chemical Society* 17, 518–520.

Du, F.Y., Xiao, X.G., Luo, X.J. and Li, G.K. (2009) Application of ionic liquids in the microwave-assisted extraction of polyphenolic compounds from medicinal plants. *Talanta* 78, 1177–1194.

Fang, Q., Yeung, H.W., Leung, H.W. and Huie, C.W. (2000) Micelle-mediated extraction and pre-concentration of ginsenosides from Chinese herbal medicine. *Journal of Chromatography A* 904, 47–55.

Fulzele, D.P. and Satdive, R.K. (2005) Comparison of techniques for the extraction of anti-cancer drug camptothecin from *Nothapodytes foetida*. *Journal of Chromatography A* 1063, 9–13.

Ganzler, K. and Salgo, Á.A. (1987) Microwave-extraction – a new method superseding traditional Soxhlet extraction. *Zeitschrift für Lebensmittel-Untersuchung und-Forschung* 184, 274–276.

Ganzler, K., Salgo, Á.A. and Valko, Â.K. (1986a) Microwave extraction. A novel sample preparation method for chromatography. *Journal of Chromatography* 371, 299–306.

Ganzler, K., Ba, Â.J. and Valko, Â.K. (1986b) A new method for the extraction and high performance liquid chromatographic determination of vicine and convicine in faba beans. *Chromatography* 84, 435–442.

Greenwald, C.G. (1991) Overview of fat and cholesterol reduction technologies. In: Haberstroh, C. and Morris, C.E. (eds) *Fat and Cholesterol Reduced Foods: Technologies and Strategies*. *Advances in Applied Biotechnology Series*, Vol. 12. Gulf Publication Company, The Woodlands, Texas, pp. 21–32.

Guclu-Ustundag, O. and Temelli, F. (2005) Solubility behavior of ternary systems of lipids, cosolvents and supercritical carbon dioxide and processing aspects. *Journal of Supercritical Fluids* 36, 1–15.

Gutiérrez, J.M.R., Jiménez, J.R. and Luque de Castro, M.D. (2008) Ultrasound-assisted dynamic extraction of valuable compounds from aromatic plants and flowers as compared with steam distillation and superheated liquid extraction. *Talanta* 75, 1369–1375.

Hao, J.Y., Han, W., Huang, S.D., Xue, B.Y. and Deng, X. (2002) Microwave-assisted extraction of artemisinin from *Artemisia annua* L. *Separation and Purification Technology* 28, 191–196.

Hartono, R., Mansoori, G.A. and Suwono, A. (2001) Prediction of solubility of biomolecules in supercritical solvents. *Chemical Engineering Science* 56, 6949–6958.

Hawthorne, S.B. and King, J.B. (1999) Principles and practice of analytical supercritical fluid. In: Caude, M. and Thiebaut, D. (eds) *Practical Supercritical Fluid Chromatography and Extraction*. Harwood Academic Publishers, The Netherlands, pp. 219–282.

Hemwimol, S., Pavasant, P. and Shotipurk, A. (2006) Ultrasound-assisted extraction of anthraquinones from roots of *Morinda citrifolia*. *Ultrasonics Sonochemistry* 13, 543–548.

Hudaib, M., Gotti, R., Pomponio, R. and Cavrni, V. (2003) Recovery evaluation of lipophilic markers from *Echinacea purpurea* roots applying microwave-assisted solvent extraction versus conventional methods. *Journal of Separation Science* 26, 97–103.

Ismail, Z. (2003) Standardization of herbal products: a case study. In a two and half day course of herbal and phytochemical processing, CEPP short course notes. Chemical Engineering Pilot Plant, UTM Skudai, 7–9 January 2003.

Jassie, L., Revesz, R., Kierstead, T., Hasty, E. and Matz, S. (1997) Microwave-assisted solvent extraction. In: Kingston H.M.S. and Haswell S.J. (eds) *Microwave-Enhanced Chemistry. Fundamentals, Sample Preparation, and Applications*. American Chemical Society, Washington, DC, pp. 569–609.

Jeong, M.L. and Chesney, D.J. (1999) Investigation of modifier effects in supercritical carbon dioxide from various solid matrices. *Journal of Supercritical Fluids* 16, 33–42.

Ju, Z.Y. and Howard, L.R. (2003) Effects of solvent and temperature on pressurized liquid extraction of anthocyanins and total phenolics from dried red grape skin. *Journal of Agricultural and Food Chemistry* 51, 5207–5213.

Karem, Z., Shashoua, H.G. and Yarden, O. (2005) Microwave-assisted extraction of bioactive saponins from chickpea (*Cicer arietinum* L.). *Journal of the Science of Food and Agriculture* 85, 406–412.

Kaufmann, B. and Christen, P. (2002) Recent extraction techniques for natural products: microwave-assisted extraction and pressurised solvent extraction. *Phytochemical Analysis* 13, 105–113.

Kawamura, F., Kikuchi, Y., Ohira, T. and Yatagai, M. (1999) Accelerated solvent extraction of paclitaxel and related compounds from the bark of *Taxus cuspidate*. *Journal of Natural Products* 62, 244–247.

Ko, T.F., Weng, Y.M. and Chiou, R.Y.Y. (2002) Squalene content and antioxidant activity of *Terminalia catappa* leaves and seeds. *Journal of Agricultural and Food Chemistry* 50, 5343–5348.

Kohler, M., Haerdi, W., Christen, P. and Veuthey, J.L. (1997) Extraction of artemisinin and artemisinic acid from *Artemisia annua* L. using supercritical carbon dioxide. *Journal of Chromatography A* 785, 353–360.

Komolpis, K., Kaufman, P. and Hang, H.Y. (1998) Chemical permeabilization and *in situ* removal of daidzein from biologically viable soybean (*Glycine max*) seeds. *Biotechnology Techniques* 12, 697–700.

Kou, D. and Mitra, S. (2003) Extraction of semivolatile organic compounds from solid matrices. In: Mitra, S. (ed.) *Sample Preparation Techniques in Analytical Chemistry*. Wiley, Hoboken, New Jersey, p. 146.

Lack, E. and Seidlitz, S. (1993) Commercial scale decaffeination of coffee and tea using supercritical carbon dioxide. In: King, M.B. and Bott, T.R. (eds) *Extraction of Natural Products Using Near Critical Solvents*. Blackie Academic and Professional, Glasgow, UK, p. 101.

Laws, D.R.J., Bath, N.A., Ennis, C.S. and Wheldon, A.G. (1980) Hop extraction with carbon dioxide. US Patent 4218491.

Lee, M.H. and Lin, C.C. (2007) Comparison of techniques for extraction of isoflavones from the root of *Radix puerariae*: ultrasonic and pressurized solvent extractions. *Journal of Food Chemistry* 105, 223–228.

Lehotay, S.J. (1997) Supercritical fluid extraction of pesticides in foods. *Journal of Chromatography A* 785, 289–312.

Letellier, M., Budzinski, H. and Garrigues, P. (1999) Focussed microwave-assisted extraction of polycyclic aromatic hydrocarbons. *LC-GC International* 12, 222–225.

Li, H., Chen, B. and Yao, S. (2005) Application of ultrasonic technique for extracting chlorogenic acid from *Eucommia ulmoides* Oliv. (*E. ulmoides*). *Ultrasonics Sonochemistry* 12, 295–300.

Liazid, A., Palma, M., Brigui, J. and Barroso, C.G. (2007) Investigation on phenolic compounds stability during microwave-assisted extraction. *Journal of Chromatography A* 1140, 29–34.

Lin, M.C., Tsai, M.J. and Wen, K.C. (1999) Supercritical fluid extraction of flavonoids from *Scutellariae radix*. *Journal of Chromatography A* 830, 387–395.

Ling, J.Y., Zhang, G.Y., Cui, Z.J. and Zhang, C.K. (2007) Supercritical fluid extraction of quinolizidine alkaloids from *Sophora flavescens* Ait. and purification by high-speed countercurrent chromatography. *Journal of Chromatography A* 1145, 123–127.

Liu, B., Lockwood, B. and Gifford, L.A. (1995) Supercritical fluid extraction of diosgenin from tubers of *Dioscorea nipponica*. *Journal of Chromatography A* 690, 250–253.

Luque de Castro, M.D. and García-Ayuso, L.E. (1998) Soxhlet extraction of solid materials: an outdated technique with a promising innovative future. *Analytica Chimica Acta* 369, 1–10.

Luque de Castro, M.D., Valcarcel, M. and Tena, M.T. (1994) *Analytical Supercritical Fluid Extraction*. Springer, Berlin.

Luthria, D.L., Mukhopadhaya, S. and Krizek, D.T. (2006) Content of total phenolics and phenolic acids in tomato (*Lycopersicon esculentum* Mill.) fruits as influenced by cultivar and solar UV radiation. *Journal of Food Composition and Analysis* 19, 771–777.

Mandal, V., Mohan, Y. and Hemalatha, S. (2007) Microvave assisted extraction – an innovative and promising extraction tool for medicinal plant research. *Pharmacognosy Reviews* 1, 7–18.

Macias-Sanchez, M.D., Mantell, C., Rodriguez, M., de la Ossa, E.M., Lubian, L.M., *et al.* (2005) Super critical fluid extraction of carotenoids and chlorophyll from a *Nannochloropsis gaditana. Journal of Food Engineering* 66, 245–251.

Martinelli, E., Schulz, K. and Mansoori, G.A. (1991) Supercritical fluid extraction/retrograde condensation with applications in biotechnology. In: Bruno, T.J. and Fly, J.F. (eds) *Supercritical Fluid Technology*. CRC Press, Boca Raton, Florida, pp. 451–478.

Mason, T.J., Paniwnyk, L. and Lorimer, J.P. (1996) The uses of ultrasound in food technology. *Ultrasonics and Sonochemistry* 3, S253–S260.

Matthaus, B. and Bruhl, L. (2001) Comparison of different methods for the determination of the oil content of oilseeds. *Journal of American Oil Chemists Society* 78, 95–102.

Mattina, M.J., Berger, W.A.L. and Denson, C.L. (1997) Microwave-assisted extraction of taxnes from *Taxus* biomass. *Journal of Agricultural and Food Chemistry* 45, 4691–4696.

Menaker, A., Kravetes, M., Koel, M. and Orav, A. (2004) Identification and characterization of supercritical fluid extracts from herbs. *Comptes Rendus Chimie* 7, 629–633.

Modey, W.K., Mulholland, D.A. and Raynor, M.W. (1996) Analytical supercritical fluid extraction of natural products. *Phytochemical Analysis* 7, 1–15.

Moore, W.N. and Taylor, L.T. (1996) Extraction and quantitation of digoxin and acetyldigoxin from the *Digitalis lanata* leaf via near super critical methanol-modified carbon dioxide. *Journal of Natural Products* 59, 690–693.

Mukhopadhyay, S., Luthria, D.L. and Robbins, R.J. (2006) Optimization of extraction process for phenolic acids from black cohosh (*Cimicifuga racemosa*) by pressurized liquid extraction. *Journal of the Science of Food and Agriculture* 86, 156–162.

Mushtaq, M.Y., Choi, Y.H., Verpoorte, R. and Wilson, E.G. (2014) Extraction for metabolomics: access to the metabolome. *Phytochemical Analysis* 25, 291–306.

Novak, I., Janeiro, P., Seruga, M. and Oliveira-Brett, A.M. (2008) Ultrasound extracted flavon-oids from four varieties of Portuguese red grape skins determined by reverse-phase high performance liquid chromatography with electrochemical detection. *Analytica Chimica Acta* 630, 107–115.

Octave, S. and Thomas, D. (2009) Biorefinery: toward an industrial mechanism. *Biochimie* 91, 659–664.

Onuska, F.I. and Terry, K.A. (1993) Extraction of pesticides from sediments using a microwave technique. *Chromatographia* 36, 191–194.

Ozer, E.O., Platin, S., Akman, U. and Hortascsu, O. (1996) Supercritical carbon dioxide extraction of spearmint oil from mint plant leaves. *Canadian Journal of Chemical Engineering* 74, 920–928.

Pan, W.H.T., Chin Chang, C., Tsu Su, T., Lee, F. and Steve Fuh, M.R. (1995) Preparative super-critical fluid extraction of pyrethrin I and II from pyrethrum flower. *Talanta* 42, 1745–1749.

Pan, X., Liu, H., Jia, G. and Shu, Y.Y. (2000) Microwave-assisted extraction of glycyrrhizic acid from licorice root. *Biochemical Engineering Journal* 5, 173–177.

Pare, J.R.J. (1990) Microwave extraction of volatile oils and apparatus therefore. Patent No. 90250286 (0 485 668 A1).

Pare, J.R.J. (1991) Microwave assisted natural products extraction. Patent No. 519 588 (5,002,784).

Pascual-Marti, M.C., Salvador, A., Chafer, A. and Berna, A. (2001) Supercritical fluid extraction of resveratrol from grape skin of *Vitis vinifera* and determination by HPLC. *Talanta* 54, 735–740.

Perez-Serradila, J.A., Priego-Capote, F. and Luque de Castro, M.D. (2007) Simultaneous ultrasound-assisted emulsification-extraction of polar and nonpolar compounds from solid plant extracts. *Analytical Chemistry* 79, 6767–6774.

Phelps, C.J., Smart, N.G. and Wai, C.M. (1996) Past, present and possible future applications of supercritical fluid extraction technology. *Journal of Chemical Education* 73, 1163–1168.

Popovich, D.G. and Kitts, D.D. (2004) Generation of ginsenosides Rg3 and Rh2 from North American ginseng. *Phytochemistry* 65, 337–344.

Pourmortazavi, S.M. and Hajimirsadeghi, S.S. (2007) Supercritical fluid extraction in plant essential and volatile oil analysis. *Journal of Chromatography A* 1163, 2–24.

Pourmortazavi, S.M., Sefidkon, F. and Hosseini, S.G. (2003) Supercritical carbon dioxide extraction of essential oils from *Perovskia atriplicifolia* Benth. *Journal of Agricultural and Food Chemistry* 51, 5414–5419.

Prasad, M.R. and Ferenci, T. (2003) Global metabolites analysis: the influence of extraction methodology on metabolome profiles of *Escherichia coli*. *Analytical Biochemistry* 313, 145–154.

Presti, M.L., Ragusa, S., Trozzi, A., Dugo, P., Visinoni, F., *et al.* (2005) A comparison between different techniques for the isolation of rosemary essential oil. *Journal of Separation Science* 28, 273–280.

Qu, H.B., Ou, D.L. and Cheng, Y.Y. (2006) A new quality control method of Chinese medicinal plant extracts. *Chinese Pharmaceutical Journal* 41, 57–60.

Quan, C., Sun, Y. and Qu, J. (2009) Ultrasonic extraction of ferulic acid from *Angelica sinensis*. *Canadian Journal of Chemical Engineering* 87, 562–567.

Reighard, T.S. and Olesik, S.V. (1996) Comparison of supercritical fluids and enhanced-fluidity liquids for the extraction of phenolic pollutants from house dust. *Analytical Chemistry* 68, 3612–3621.

Renoe, B.W. (1994) Microwave assisted extraction. *American Laboratory* 26, 34–40.

Richter, B.E., John, B.A., Ezzell, L., Porter, N.L., Avdalovic, N. and Pohl, C. (1996) Accelerated solvent extraction: a technique for sample preparation. *Analytical Chemistry* 68, 1033–1039.

Richter, P., Jimenez, M., Salazar, R. and Marican, A. (2006) Ultrasound-assisted pressurized solvent extraction for aliphatic and polycyclic aromatic hydrocarbons from soils. *Journal of Chromatography A* 1132, 15–20.

Rompp, H., Seger, C., Kaiser, C.S., Haslinger, E. and Schmidt, P.C. (2004) Enrichment of hyperforin from St. John Wort (*Hypericum perforatum*) by pilot-scale super critical carbon dioxide extraction. *European Journal of Pharmaceutical Science* 21, 443–451.

Rouhani, S., Alizadeh, N., Salimi, S. and Haji-Ghasemi, T. (2009) Ultrasonic assisted extraction of natural pigments from rhizomes of *Curcuma longa* L. *Progress in Colours, Colourants and Coatings* 2, 103–113.

Sargenti, S.R. and Vichnewski, W. (2000) Sonication and liquid chromatography as a rapid technique for extraction and fractionation of plant material. *Phytochemical Analysis* 11, 69–73.

Scalia, S., Giuffreda, L. and Pallado, P. (1999) Analytical and preparative supercritical fluid extraction of chamomile flowers and its comparison with conventional methods. *Journal of Pharmaceutical and Biomedical Analysis* 21, 549–558.

Schafer, K. (1998) Accelerated solvent extraction of lipids for determining the fatty acid composition of biological material. *Analytica Chimica Acta* 358, 69–77.

Schinor, E.C., Salvador, M.J. and Turatti, I.C.C. (2004) Comparison of classical and ultrasound-assisted extractions of steroids and triterpenoids from three *Chresta* spp. *Ultrasonics Sonochemistry* 11, 415–421.

Shu, S., Gao, Z.H. and Yang, X. (2004) Supercritical fluid extraction of sapogenins from tubers of *Smilax China*. *Fitoterapia* 75, 656–661.

Shu, Y.Y., Ko, M.Y. and Chang, Y.S. (2005) Microwave assisted extraction of ginsenosides from ginseng root. *Michrochemical Journal* 74, 131–139.

Silas, G. and Villas, B. (2006) Sampling and sample preparations. In: Villas, B., Silas, G., Ute, R., Hansen, M.A.E., Smedsgaard, J. and Nielsen, J. (eds) *Metabolome Analysis: An Introduction*, Vol 1. Wiley, Hoboken, New Jersey, pp. 39–76.

Simandi, B., Kristo, S.T., Kery, A., Selmeczi, L.K. and Kemeny, S. (2002) Supercritical fluid extraction of dandelion leaves. *The Journal of Supercritical Fluids* 23, 135–142.

Smith, R.M. and Burford, M.D. (1992a) Supercritical fluid extraction and gas chromatographic determination of the sesquiterpene lactone parthenoline in the medicinal herb feverfew (*Tanacetum parthenium*). *Journal of Chromatography A* 627, 255–261.

Smith, R.M. and Burford, M.D. (1992b) Optimization of supercritical fluid extraction of volatile constituents from a model plant matrix. *Journal of Chromatography* 600, 175–181.

Song, K.M., Park, S.W., Hong, W.H., Lee, H., Kwak, S.S., *et al.* (1992) Isolation of vindoline from *Catharanthus roseus* by supercritical fluid extraction. *Biotechnology Progress* 8, 583–586.

Sovová, H., Zarevúcka, M., Vacek, M. and Stránsky, K. (2001) Solubility of two vegetable oils in supercritical CO_2. *Journal of Supercritical Fluids* 20, 15–28.

Spiro, M. and Chen, S.S. (1995) Kinetics of isothermal and microwave extraction of essential oils constituents of peppermint leaves into several solvent systems. *Flavour and Fragrance Journal* 10, 259–272.

Stavarache, C., Vinatoru, M., Nishimura, R. and Maeda, Y. (2005) Fatty acids methyl esters from vegetable oil by means of ultrasonic energy. *Ultrasonics Sonochemistry* 12, 367–376.

Sun, L., Rezaei, K.A., Temelli, F. and Ooraikul, B. (2002) Supercritical fluid extraction of alkylamides from *Echinacea angustifolia*. *Journal of Agricultural and Food Chemistry* 50, 3947–3953.

Synder, L.R. (1978) Classification of the solvent properties of common liquids. *Journal of Chromatographic Science* 16, 223–224.

Tatke, P. and Jaiswal, Y. (2011) An overview of microwave assisted extraction and its applications in herbal drug research. *Research Journal of Medicinal Plants* 5, 21–31.

Taylor, L.T. (1996) *Supercritical Fluid Extraction*. Wiley, Chichester, UK.

Tigrine-Kordjani, N., Meklati, B.Y. and Chemat, F. (2006) Microwave dry distillation as a useful tool for extraction of edible essential oils. *International Journal Aromatherapy* 16, 141–147.

Toma, M., Vinatoru, M., Paniwnyk, L. and Masson, T.J. (2001) Investigation of the effects of ultrasound on vegetal tissues during solvent extraction. *Ultrasonics Sonochemistry* 8, 137–142.

Tonthubthimthong, P., Douglas, P.L., Douglas, S., Luewisutthicat, W., Teppaitoon, W., *et al.* (2004) Extraction of nimbin from neem seeds using supercritical carbon dioxide and supercritical carbon dioxide-methanol mixture. *The Journal of Supercritical Fluids* 30, 287–301.

Turner, C., Eskilsson, C.S. and Bjorklund, E. (2002) Collection in analytical scale supercritical fluid extraction. *Journal of Chromatography A* 947, 1–22.

Vasapollo, G., Longo, L., Rescio, L. and Ciurlia, L. (2004) Innovative supercritical CO_2 extraction of lycopene from tomato in the presence of vegetable oil as co-solvent. *The Journal of Supercritical Fluids* 29, 87–96.

Vilegas, J.H.Y., de Marchi, E. and Lancas, F.M. (1997) Extraction of low-polarity compounds (with emphasis on coumarin and kaurenoic acid) from *Mikania glomerata* ('guaco') leaves. *Phytochemical Analysis* 8, 266–270.

Vinatoru, M., Toma, M., Radu, O., Filip, P.I., Lazurca, D., *et al.* (1997) The use of ultrasound for the extraction of bioactive principles from plant materials. *Ultrasonics Sonochemistry* 4, 135–139.

Waksmundzka-Hajnos, M., Petruczynik, A., Dragan, A., Wianowska, D., Dawidowicz, A.L., *et al.* (2004) Influence of the extraction mode on the yield of some furanocoumarins from *Pastinaca sativa* fruits. *Journal of Chromatography B* 800, 181–187.

Wang, H.C., Chen, C.R. and Chang, C.J. (2001b) Carbon dioxide extraction of ginseng root hair oil and ginsenosides. *Food Chemistry* 72, 505–509.

Wang, H.Y., Komolpis, K., Kaufman, P.B., Malakul, P. and Shotipruk, A. (2001a) Permeabilization of metabolites from biologically viable soybeans (*Glycine* max.). *Biotechnology Progress* 17, 424–430.

Warburton, E., Norris, P. and Goenaga-Infante, H. (2007) Comparison of the capabilities of accelerated solvent extraction and sonication as extraction techniques for the quantification of kavalactones in *Piper methysticum* (Kava) roots by high performance liquid chromatography with ultra violet detection. *Phytochemical Analysis* 18, 98–102.

Wijngaard, H. and Brunton, N. (2009) The optimization of extraction of antioxidants from apple pomace by pressurized liquids. *Journal of Agricultural and Food Chemistry* 57, 10625–10631.

Wright, B.W., Wright, C.B., Gale, R.W. and Smith, R.D. (1987) Analytical supercritical fluid extraction of adsorbent materials. *Analytical Chemistry* 59, 38–44.

Wu, J., I in, L. and Chau, F.T. (2001) Ultrasound-assisted extraction of ginseng saponins from ginseng roots and cultured ginseng cells. *Ultrasonics Sonochemistry* 8, 347–352.

Xing, H., Yang, Y., Su, B., Huang, M. and Ren, Q. (2003) Solubility of artemisinin in supercritical carbon dioxide. *Journal of Chemical Engineering Data* 48, 330–332.

Yamini, Y., Asghari-Khiavi, M. and Bahramifar, N. (2002) Effects of different parameters on supercritical fluid extraction of steroid drugs from spiked matrices and tablets. *Talanta* 58, 1003–1010.

Yang, L., Wang, H., Zu, Y.G., Zhao, C.J., Zhang, L., *et al.* (2011) Ultrasound-assisted extraction of the three terpenoid indole alkaloids vindoline, catharanthine and vinblastine from *Catharanthus roseus* using ionic liquid aqueous solutions. *Chemical Engineering Journal* 172, 705–712.

Yoda, S.K., Marques, M.O.M., Petenate, A. and Meireless, M.A.M. (2003) Supercritical fluid extraction from *Stevia rebaudiana* Bertoni using carbon dioxide and carbon dioxide and water: extraction kinetics and identification of extracted components. *Journal of Food Engineering* 57, 125–134.

Young, J.C. (1995) Microwave assisted extraction of the fungal metabolite ergosterol and total fatty acids. *Journal of Agricultural and Food Chemistry* 43, 2904–2910.

Zhao, H., Xia, S. and Ma, P. (2005) Use of ionic liquids as green solvents for extractions. *Journal of Chemical Technology and Biotechnology* 80, 1089–1096.

Ziemons, E., Goffin, E., Lejeune, R., da Cunha, P., Angenot, L., *et al.* (2005) Supercritical carbon dioxide extraction of tagitinin C from *Tithonia diversifolia*. *The Journal of Supercritical Fluids* 33, 53–60.

Zosel, K. (1981) Process for decaffeination of coffee. 1981 US patent no. 4247570.

3 Supercritical Fluid Extraction

3.1 Introduction

Depending on temperature and pressure, substances can transform into three different states (gas, liquid and solid). Three phases of gaseous, liquid and solid coexist at the triple point. However, there is a region of pressure and temperature above this critical point to which neither liquid nor gas belongs. A supercritical fluid is liquid that is above its critical temperature and pressure (Bravi *et al.*, 2007).

It has the unique characteristics of being a solvent that continuously varies in solubility. Charles Cagniard first discovered supercritical fluids (SCFs) in 1822 and their high solvation power was first reported over a century ago. Demonstration of SFE technology for industrial applications was reported by Zosel at the Max Planck Institute für Kohlenforschung, Germany in 1969 (Zosel, 1969).

SCFs have or display properties of both gases and liquids, diffusing into a solid like a gas while dissolving like liquids. Some of the physical properties of gas, liquid and SCFs are summarized in Table 3.1.

The development of new isolation and fractionation methods for various raw materials, as well as the optimization of extraction parameters, is an important aspect of natural product research. As the biologically active compounds in plants are usually low in concentration, a great deal of research has been done to develop more effective and selective extraction methods to recover these compounds from raw materials. Further, such methods should be technologically efficient, safe and environmentally friendly, particularly when the products are intended for food and pharmaceutical purposes. Conventional solvent extraction methods do not meet these requirements. Also, the conventional extraction

 © Satyanshu Kumar 2016. *Analytical Techniques for Natural Product Research* (S. Kumar)

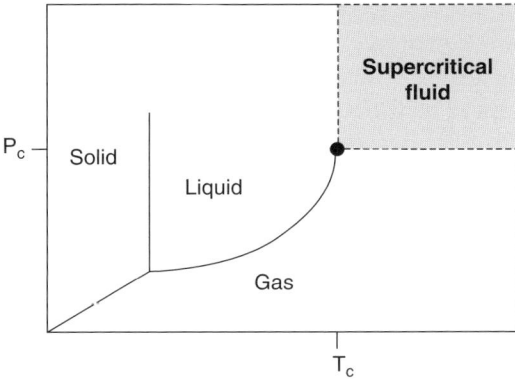

Fig. 3.1. Phase diagram for a single substance. P_c = critical pressure; T_c = critical temperature.

Table 3.1. Physical properties of gas, liquid and supercritical fluid.

	Diffusivity DM (cm^2 s^{-1})	Density ρ (g cm^{-3})	Viscosity (g cm^{-1} s^{-1})
Gas	10^{-1}	10^{-3}	10^{-4}
Supercritical fluid	10^{-3}	0.2–0.8	10^{-3}
Liquid	10^{-6}	1	10^{-2}

methods used to obtain these types of products have several drawbacks, such as:

1. They are time-consuming.
2. They are laborious in nature.
3. They have low selectivity.
4. Extraction yield is low.

Also, for conventional methods like solvent extraction and hydrodistillation, there are few adjustable parameters to control the selectivity of the extraction process. Moreover, these traditional techniques employ large volumes of toxic solvents. Also, with increasing legal restrictions reflecting consumers' concern regarding the use of organic solvents, extraction alternatives are required for additive-free natural products. Therefore, developing alternative extraction techniques with better selectivity and efficiency is highly desirable. Extraction techniques that are non-aggressive to the environment and that produce products and subproducts free of undesirable solvent residues have been suggested as alternatives to conventional processes. Some of the organic solvents, such as methylene chloride, have induced health concerns due to environmental emissions and/or trace residues left in the product. This has propelled research efforts aimed at developing environmentally benign processing techniques. The employment of such a technique either eliminates or reduces significantly the environmental problem. At present, extraction methods able to overcome the above-mentioned drawbacks are being studied, and among them one method is supercritical fluid extraction (SFE) (Subramaniam *et al.*, 1993). SFE is defined as the separation of one component from others with the use of an SCF. SFE has received much attention in past decades, especially in the food, pharmaceutical and cosmetic industries, because it presents an alternative to conventional processes such as organic solvent extraction, steam distillation and low-temperature separation, and also prevents the degradation of chemical compounds (Sovova, 2005). The combined liquid-like solvating capabilities and gas-like transport properties of SCFs make them particularly suitable for the extraction of diffusion-controlled matrices such as plant tissues. Also, a small change in temperature and pressure can change the properties of SCFs dramatically, thereby, making it possible to control their physico-chemical properties such as density and solvation power (Abbas *et al.*, 2008). SCFs are non-explosive, have low viscosity, high diffusivity and low toxicity, as well as low surface tension, which could improve the extraction yield of desired materials from complex matrices. Also, the solvent strength of an SCF can be adjusted easily by simply changing the applied pressure and temperature (Casas *et al.*, 2008).

The extraction of phytochemicals using SCFs is of great importance, due to the high purity of the final products and their price in the international market. Several solvents are used for SFE; however, carbon dioxide is an ideal solvent. The most widely used SCF is carbon dioxide, and more than 90% of all analytical supercritical extraction is performed with supercritical carbon dioxide (SC-CO_2), for several practical reasons. Apart from relatively low critical pressure and temperature (73.8 bars, 31.1°C) and high purity, it is additionally almost inert in standard operating conations. SC-CO_2 is non-toxic, non-inflammable, available at relatively low cost and is removed easily from the extract (Pasquali *et al.*, 2006). As chemical reactions do not take place during the extraction in SFE, it is largely ensured that isolated analytes are representative of the original sample.

The potential advantages of SFE accrue from the properties of a solvent at temperature and pressure above its critical point. At elevated pressure, this single phase will have properties that are intermediate between those of the gas and liquid phases and are dependent on the fluid composition, pressure and temperature. The compressibility of SCF is large just above the critical temperature and small changes in pressure result in large changes in the density of the fluid. The density of SCF is typically 100–1000 times greater than that of the gas. Consequently, molecular interaction increases due to shorter intermolecular distance. Diffusion coefficients and viscosity of the fluid, although density dependent, remain more similar to those of a gas. The liquid-like behaviour of an SCF results in greatly enhanced solubilizing capabilities of the corresponding liquid. These properties allow similar solvent strength to that of liquids, but with greatly improved mass transfer properties that provide the potential for more rapid extraction rates and more efficient extraction due to better penetration of the matrix.

3.2 Supercritical Fluid as a Solvent

Any solvent can be used as a supercritical solvent depending on its technical viability in terms of critical properties, toxicity, cost and solvation capacity. Besides SC-CO$_2$, other solvents such as propane, ethane, hexane, pentane and butane have also been investigated as an SCF. In spite of the toxicity of certain solvents, their use has been advocated because at or near supercritical conditions the amount of solvent used is much smaller than the required amount for any extraction process at low pressure. The critical properties of some of the SCFs are listed in Table 3.2.

SFE can be carried out at low temperatures, leading to less deterioration of the thermally labile compounds in the extract, with comparable or better recovery and minimal alteration of the active ingredients; thus, the curative properties can be preserved. So, SFE is widely used to extract and separate active compounds in the food, pharmaceutical and cosmetic industries (Mostafa *et al.*, 2004).

3.3 Optimization of Extraction Parameters

One of the main aspects of SFE is to recognize the best extraction conditions. Using appropriate values for different variables influencing SFE, recovery of the extraction yield of target compounds could be improved significantly. In comparison to other extraction techniques, the optimization of SFE is a complex process due to the involvement of a multitude of

Table 3.2. Critical properties of some supercritical fluids.

Solvent	Critical temperature (°C)	Critical pressure (bars)	Critical volume (cm^3 mol^{-1})	Reference
Ammonia	132.4	113.5	72.5	Jacobson *et al.*, 1997
Carbon dioxide	31.1	73.7	94.1	Pereira and Meireles, 2010
Dimethyl ether	127.1	52.7	171.0	Catchpole *et al.*, 2003
Ethane	32.3	48.7	145.5	Mohamed *et al.*, 2002
Ethylene	9.3	50.4	131.0	Shende and Lombardo, 2002
Methanol	239.6	80.9	118.0	Capriel *et al.*, 1986
n-Hexane	234.5	30.2	368.0	Joyce and Thies, 1997
Propane	96.8	42.5	200.0	Catchpole *et al.*, 2003
Water	374.1	220.6	55.9	Leal *et al.*, 2008
Xenon	17.1	58.0	118.0	Liang and Tilotta, 1998

parameters such as extraction time, pressure, temperature, flow rate of SCF and modifier or co-solvent percentage. In addition to the large number of parameters, each factor can also have a marked effect on extraction efficiency (Casas *et al.*, 2009). Often, SFE methods involve the investigation of many variables that may affect extraction efficiency. The selection of these variables and their levels is critical. Extraction pressure, temperature of the fluid and extraction time are generally considered as the most important factors, as these parameters potentially affect the extraction process (Wakte *et al.*, 2011). Since various parameters potentially affect the extraction process, optimization of the experimental conditions represents a critical step in the method development of an SFE process. There have been considerable research efforts concerning the effects of different parameters such as temperature, pressure, sample matrix, extraction time and addition of modifier on the extraction behaviour of analytes. Therefore, the first step in SFE is to optimize the operating conditions to obtain an efficient extraction of the target compounds and to avoid the co-extraction of undesired compounds.

Optimization of the method can be carried out step by step or by using an experimental design. Several statistical methods such as simplex optimization and full factorial design are employed for the optimization of analytical SFE methods. Optimization of the extraction procedure by full factorial design is faster and more economical and effective than traditional one at a time (Machmudah *et al.*, 2007, 2008; Solati *et al.*, 2012). Full factorial optimization makes it possible to understand circumstances that are not explained by the traditional approach; for example, the interactions between the factors that influence analytical responses (Bezerra *et al.*, 2008). Maximum or minimum levels of response functions of critical experiment conditions are assessed using this model of design. Although full factorial designs require more experimental points, it could provide more information in terms of the effect of variable combinations. Response surface methodology (RSM) is used to evaluate the effect of multiple factors and their interaction on one or more response variables. RSM uses the multiple regression model (a polynomial, second-order equation). RSM involves a set of empirical techniques for finding the relative influence of the controlled independent factors such as temperature, pressure, extraction time and flow rate of SCF. It evaluates the quantitative relationship between the set of input variables and the measured responses (dependent variable) according to one or more selected criteria (Teruel *et al.*, 1997; Sanal *et al.*, 2005). Moreover, it can be used to determine the optimum levels of input variables required for a given response. Since it defines the interactions between input variables, repeated runs are reduced. However, a prior knowledge and understanding of the process and the process variables under investigation are necessary to achieve a realistic model. Central composite design (CCD) is the most popular form of RSM, and it could be utilized to optimize the extraction methods. As compared to full factorial designs, RSM requires fewer experimental runs (Tan *et al.*, 2009). All the selected factors are examined using an orthogonal design, and in order to determine a suitable extraction condition in a wide range with a minimum number of trials, an orthogonal test design is employed where temperature, pressure and time are considered to be the three major factors for effective extraction (Chen *et al.*, 2009). The best way to visualize the effect of independent variables on a dependent variable is to draw surface response plots of the model. The relationship between the responses and the experimental variables can be illustrated graphically to investigate the interactions of the variables and to determine the optimum level of each variable for the maximum response by plotting three-dimensional response surface plots. Each plot shows a pair of factors by keeping the other factor constant at its middle level. RSM involves three steps: (i) central composite rotatable design; (ii) response surface modelling through regression analysis; and (iii) process factor optimization using the response surface models.

The solubility of target compound in SCF is a major factor determining its extraction efficiency. The solubility of solutes in SC-CO_2 is affected by two competing factors in opposing ways and is controlled by the sum of the volatility of the substance, which is temperature dependent, and the solvation effect of the SCF, which is density dependent (Modey *et al.*, 1996). The density of supercritical carbon dioxide and the volatility of the solute depend on the temperature. Higher temperatures increase the volatility of solutes and improve their solubility and extraction. On the other hand, at a given pressure, the density of supercritical carbon dioxide decreases with increasing temperature, because of the increase in vapour pressure, thereby reducing the solvating power of carbon dioxide and thus reducing solubility and extraction efficiency (Solati *et al.*, 2012). Phelps *et al.* (1996) suggested that changing both the temperature and the pressure of fluids could modify the solvating power of organic solvents. As a rule, an increase in temperature at constant pressure reduces solvating density and its solvating power, whereas, an increase of pressure at constant temperature enhances solvation power (Pereira and Meireles, 2010). The number of polar functional groups, for example hydroxyl groups, may affect the volatility of the solutes, thereby determining extractability with SC-CO_2. Stahl and Glatz (1984) were able to extract steroids with three functional groups below 300 bars, but they were unable to extract those steroids containing four hydroxyl groups or three hydroxyl groups and one acid group or one phenolic hydroxyl with two other hydroxyl groups. For SFE of β-carotene from paprika, Weathers *et al.* (1999) observed that when pressure was reduced from 338 to 250 bars, which corresponded to a decrease in fluid density, extraction efficiency was reduced dramatically. Langenfeld *et al.* (1993) indicated that an increase in extraction efficiency could be dependent on the molecular weight, which was related to the vapour pressure of the molecule. Also, the solubility of solid compounds in SCF could be influenced by a repulsive solute–fluid interaction (Ma *et al.*, 2008). Increasing the extraction pressure at fixed temperature results in an increase in the fluid density and solvating power of SC-CO_2 (Oostdyk *et al.*, 1993), which in many cases improves extraction efficiencies (Ashraf-Khorassani *et al.*, 1995; Daneshfar *et al.*, 1995). Generally, increase in pressure could result in the enhanced extraction of polar and high molecular mass compounds such as polyphenols, and when no high molecular weight compounds are present in the matrix, the increase could not enhance extraction yield (Esquivel *et al.*, 1999). The effect of temperature and static time on extraction efficiency, however, has proven to be analyte and matrix dependent. Although increasing the temperature at fixed pressure lowers fluid density in certain cases, extraction efficiency was reported to be enhanced by an increase in temperature at fixed pressure (Berger, 1989; Hutz *et al.*, 1990). The variation of temperature during the SFE process affects the density of the fluid, the volatile property of the analytes and desorption of the analytes from the matrix. At higher temperature, the analytes become more volatile but the density of SC-CO_2 decreases. Hawthorne *et al.* (1993) reported that an increase in extraction efficiency with increase in temperature could be dependent on the molecular weight, which is related to vapour pressure. Increasing the dynamic extraction time is usually found to increase the recovery of analytes. However, this increase is not linear and curves become convex and flat after a time. In some cases, higher pressures are required for the complete recovery of the main components of the plant. This is because raising the extraction pressure at constant temperature leads to higher fluid density, which increases the solubility of the analytes. At higher pressure, extraction yield increased with increase in temperature. But at lower pressure, the inverse behaviour was observed, with extraction yield decreasing with increasing temperature at constant pressure (Filho *et al.*, 2008). It is more difficult to predict the influence of temperature on extraction than that of pressure as temperature has two opposing effects (Wang *et al.*, 2012).

Solvent flow rate is also an important factor of SFE. Generally, an increase in flow rate increases extraction capacity, but in some cases very high flow rates can decrease it. Mass transfer coefficients increase with an increase in the Reynolds number, and higher flow rates give larger mean concentration gradients because the loading of the extraction solvent with a solute is lower and better mixing affects extraction (Sovilj *et al.*, 2011). Increase of flow rate increased the oil yield of black pepper (Perakis *et al.*, 2005).

The effect of solvent flow rate on the extraction of andrographolide from the leaves of *Andrographis paniculata* was reported by Kumoro and Hasan (2007). It was observed that very high flow rates resulted in lower extraction efficiency. Insufficient contact time between the solvent and the sample was suggested as the probable reason, considering the low diffusion rate of the solvent–solute within the matrix (Mushtaq *et al.*, 2014). Nemoto *et al.* (1997) studied the effect of the CO_2 flow rate on the extraction of pesticide, and it was proposed that low flow rates resulted in low linear velocity and increase in extraction efficiency.

3.3.1   Limitations of SC-CO_2 and modifier or co-solvent

The main disadvantage of SC-CO_2 as a solvent is its non-polar nature. It has good solvent properties for the extraction of non-polar compounds, and a large quadrupole moment also enables it to dissolve some moderately polar compounds like alcohols, esters, aldehydes and ketones (Lang and Wai, 2001). However, this can be overcome by adding modifiers or co-solvents. A modifier can be added with the solvent or together with the sample in the extraction chamber. The first method can be applied to both countercurrent SFE and static SFE. The second method is always used for static SFE. These modifiers can range from ethanol, acetonitrile and methanol to water. It is assumed that the effect of a modifier depends on the nature of solute to be extracted (Walsh *et al.*, 1987). The first basis of modifier selection is the increased solubility of the target compounds in the modified SCF (Pourmortazavi and Hajimirsadeghi, 2007). The ability to distort and swell the matrix as a consequence, favouring the penetration of CO_2 into the matrix for extracting the analyte, is another basis for selecting the modifier (Casas *et al.*, 2007). The effect of the modifier is determined by the phase behaviour of the mixture under operating conditions and is dependent on the concentration of co-solvent in the supercritical phase (Shi *et al.*, 2009). At least 17 modifiers have been studied in the SFE of natural products (Modey *et al.*, 1996). The selection of a modifier becomes a great challenge. Water and ethanol become attractive alternatives over other organic-based solvents because of their relatively safe nature and the advantages of cleanness (Shi *et al.*, 2009). But, in practice, methanol and acetone are the most popular among the different modifiers (Engelhaqrdt and Hass, 1993; Taylor *et al.*, 1993; Phelps *et al.*, 1996; Jeong and Chesney, 1999). Vegetable oils have also been used as a modifier for $SC-CO_2$ (Vasapollo *et al.*, 2004; Sun and Temeli, 2006; Krichnavaruk *et al.*, 2008). Up to 20% methanol is miscible with CO_2 and it is an effective polar modifier. Ethanol, although not as polar as methanol, may be a better choice in the SFE of natural products because of its lower toxicity. Generally, their concentration ranges between 1 and 15%. The addition of a small percentage of polar organic modifiers to supercritical carbon dioxide can increase the extractability of target analytes remarkably. To summarize, modifier is added to an extraction process mainly for two reasons: (i) to increase the polarity of supercritical carbon dioxide; and (ii) to facilitate the desorption of analytes from the plant matrix (Onuska and Terry, 1989; Wright *et al.*, 1989).

When a modifier is added, viscosity also increases. The addition of a large amount of modifier will change the critical parameters of the mixture (Pourmortazavi and Hajimirsadeghi, 2007). The addition of ethanol increases the bulk density of $SC-CO_2$ due to the higher density of the modifier and clustering of $SC-CO_2$ molecules around the modifier (Guclu-Ustundag and Temeli, 2004). Water has also been used as a modifier in a number of $SC-CO_2$ applications, although water is only about 0.3% soluble in $SC-CO_2$ (Lehoty, 1997). Water could enhance the interaction of the analyte–modifier matrix because water can open pores and swell the matrix, thereby allowing the fluid better access to analytes to draw the analytes out of the matrix. Caffeine and epigallocatechin gallate were extracted from green tea using $SC-CO_2$ with water as the modifier and the yield increased significantly with an increase in water (Shi *et al.*, 2009).

Several mechanisms that might explain the activity of the modifier have been suggested. The existence of a solute–modifier interaction in which the modifier interacts chemically with the solute is one theory (Pereira and Meireles, 2010). The formation of a solvation shell around the solute is another theory (Yonker *et al.*, 1986). The polar modifier molecules accelerate desorption processes by competing with the analytes for active binding sites, as well as by disrupting the matrix structures (Turner *et al.*, 2001). A higher percentage of methanol could disrupt the bonding between solutes and matrices (Bicchi *et al.*, 1991). Vegetable oils have also been used as modifiers to increase the extraction of carotenoids and lycopenes in CO_2-SFE (Sun and Temeli, 2006; John *et al.*, 2009). Depending on the properties of the samples and desired compounds, the best modifier is usually determined based on the results of preliminary experiments. In the SFE of phenolic acids, methanol was a more effective modifier than acetonitrile, acetone, water or dichloromethane (DCM) (Ashraf-Khorassani *et al.*, 1995). In the SFE of paclitaxel and baccatin III, DCM was the most effective among methanol, ethyl acetate, DCM and diethyl ether for paclitaxel; however, diethyl ether was the best for baccatin III (Chun *et al.*, 1994). Smith and Burford (1992) reported that 4% methanol or chloroform did not result in any improvement in the recovery of santonin (a sesquiterpene lactone), but 4% acetonitrile could increase recovery from 38% to 85%, while water-saturated CO_2 could further increase recovery to 92%. Wang *et al.* (2001) compared the efficiency of (sub- and supercritical carbon dioxide) extraction with Soxhlet and ultrasound extraction of ginsenosides (Rb_1, Rb_2, Rc, Rd and Rg_1) and crude oil from ginseng root hair. Extraction with SC-CO_2 supplemented with ethanol (6 mole percentage) as the modifier produced significantly higher extraction yield than SFE without a modifier. Further, with former conditions at 31.2 MPa and 333 K, it was as effective as hot water extraction, but yield was lower when compared with a Soxhlet extraction using ethanol. Wood *et al.* (2006) tested methanol and dimethylsulfoxide as modifiers for their effect on extraction efficiency. With relatively higher modifier concentration (27–30 mole percentage) and consequently higher operating conditions (e.g. temperature, pressure), it was feasible to extract up to 90% of the total ginsenosides content, which was comparable to Soxhlet extraction with methanol. It was also reported that with increasing mole fraction of the modifier and with increasing pressure and temperature, solubility increased enormously, with increase in solubility 100–150 times more for eflucimibe with modifier (Sauceau *et al.*, 2004). It was observed that increasing the concentration of ethanol from 2 to 10% in SCF showed a 70–80 times increase in the solubility of gallic acid, (+) catechin and (–) epicatechin in grape seed extraction (Murga *et al.*, 2000). The observed modifier effects can be explained not only by density effects but by the effect of molecular interactions on the basis of compound solubility parameters.

3.3.2 SFE and molecular weight of target analyte

Extract is a complex mixture of compounds. The chemical composition of extracts varies with the operating conditions (temperature and pressure). The variation of chemical composition with pressure (solvent density) is reported extensively in the literature, and high molecular weight compounds are extracted mainly at high pressure (Reverchon, 1997; Quispe-Condori *et al.*, 2008). SFE can fulfil these requirements and can be an alternative to conventional extraction methods.

3.3.3 SFE and particle size

The SFE process is influenced by particle size, and its effectiveness and consumption of supercritical carbon dioxide depends on grinding efficiency (Zizovic *et al.*, 2007). Increase in the solvent to feed ratio is also important for improving SFE extraction efficiency. The morphology of the material and the location of the solute in the plant material also control the extraction rate. If the desired solute is on the surface of the material, generally extraction rates are high. However, when the desired compounds are deeper in the material, it takes more time to extract them. In these cases, mass transfer depends on particle shape and size and the porosity of the solid material (Oman *et al.*, 2013). Further, if the structure of the material is more complex and the desired compounds are deeper inside, a greater resistance to diffusion is expected (Gorbaty and Bondarenko, 1998; Beckman, 2004; Reverchon and De Marco, 2006; Diaz and Brignole, 2009; Pereira and Meireles, 2010). Therefore, the preparation of a sample is very important for the SFE of natural products (Weathers *et al.*, 1999; Lang and Wai, 2001). Smaller particles provide faster extraction due to lower diffusion paths and less diffusion resistance (Oman *et al.*, 2013). The influence of the particle size of *Maydis stigma* on the extraction yield of flavonoids was investigated (Liu *et al.*, 2011). When particle size was reduced, the contact area increased; hence, a higher yield was obtained. However, when the particle sizes are too small, this lowers the extraction yield due to agglomeration.

3.3.4 Sample preparation for SFE

Usually in SFE, the solid samples are mixed with proper amounts of inert material such as celite, HMX sand and glass beads and are then packed into the extraction vessel. This process prevents channelling, increases the contact surface between the sample and SCF and consequently reduces the equilibration time. The extraction of analytes from an inert matrix prior to study the extraction of real samples mixed with materials such as glass beads has been suggested (Andersen *et al.*, 1989; Furton and Rein,

1991). The elucidation of analyte extrability is also an important step in optimizing the SFE method. This is commonly done by applying the analyte to an inert support material, assuming no analyte–matrix interaction and change of the temperature and pressure of the carbon dioxide, to find the mildest possible interaction conditions (Bjorklund *et al.*, 1998).

3.3.5   SFE static and dynamic extraction time

The static period is adapted to promote greater contact between the particles and supercritical carbon dioxide. Also, a static extraction period is employed in order to increase the sample–extractant contact duration, followed by a dynamic extraction period in which extractant is passed continuously through the extraction chamber, thus shifting the equilibrium towards a quantitative approach.

3.4   SFE Procedures

A simple SFE process consists of two major steps: (i) the extraction of soluble substances by SCF; and (ii) the separation of these compounds from the SCF after expansion. The extraction process is illustrated in Fig. 3.2. Carbon dioxide is delivered from high-pressure cylinders. Liquefied carbon dioxide inside the cylinder is under the pressure corresponding with its gas–liquid equilibrium (about 57 bars at 20°C). If the cylinder is equipped with a dip tube, liquid carbon dioxide can be withdrawn without any significant pressure drop due to liquid phase evaporation. Liquid carbon dioxide is pressurized with a high-pressure pump and then charged into the extraction vessel with or without a small percentage of modifier to extract the compound of interest. Liquid modifiers are typically high-performance liquid chromatography (HPLC)-grade or analytical-grade solvents. Extracted compound is then transferred from the extraction vessel to a fraction collector. An automated back pressure regulator located between these vessels allows for controlled depressurization of the compounds of interest and carbon dioxide. After exiting the automated back pressure regulator, the system pressure is reduced to near atmospheric conditions, causing carbon dioxide to lose its solvating power. The extracted material is collected into a fraction collector and the condensed carbon dioxide is sent to vent, or it can be recycled to the desired pressure. The extraction yield (mass of extract/mass of dry matter) is used as an indicator of the effects of the extraction conditions.

Wide ranges of applications have been reported for SFE with SC-CO_2 in extraction and purification processes, and this technology has been used to extract bioactive compounds from natural products for a long time. Countercurrent SFE has been applied to modify essential oils; for example, the deterpenation of orange peel oils (Espinosa *et al.*, 2005) or

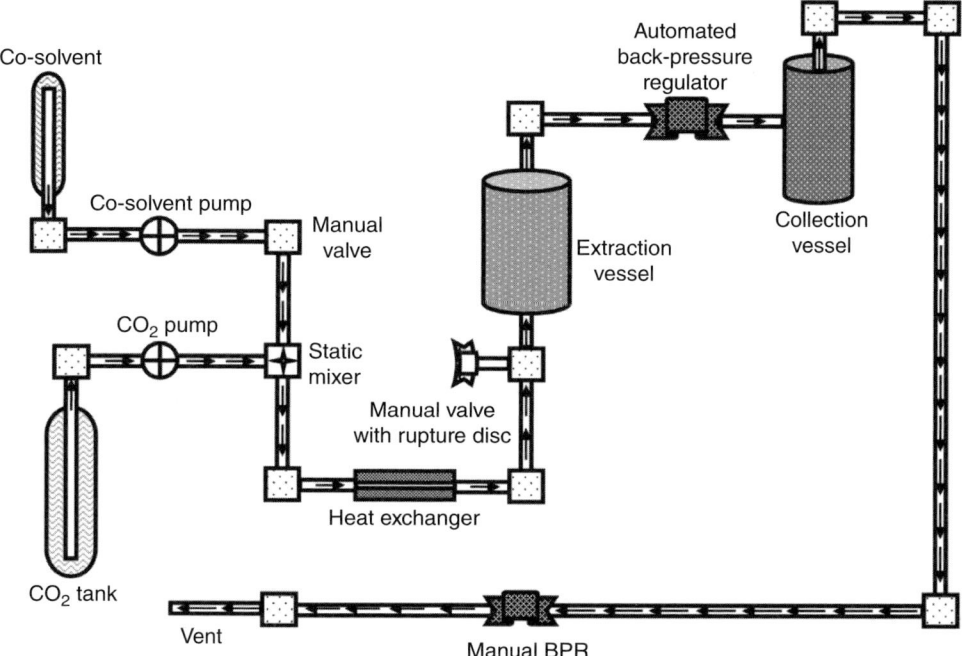

Fig. 3.2. Schematic diagram of the SFE process at the Directorate of Medicinal and Aromatic Plants Research, Anand, India.

obtaining high-value products from waste such as purification of squalene from vegetable oils (Vazquez *et al.*, 2007). Decaffenation of coffee beans with SFE is one example of extraction through solid particles or bed.

3.5 Comparison of SFE with other Extraction Techniques

SFE is a powerful alternative to conventional organic solvent extraction because of its combination of gas-like mass transfer and liquid-like solvating properties. Analytical scale SFE has been shown to be a rapid and quantitative method for extracting relatively non-polar components from a variety of sample matrices. For moderately polar compounds, SFE can be applied at higher densities or with the addition of organic modifiers. Studies have shown that for relatively polar phenolic compounds supercritical carbon dioxide is not a good choice to give high extraction yields (Tena *et al.*, 1997). Interaction of the target analytes with the sample matrix could have a major effect on extraction efficiency. The effects of different parameters such as pressure, temperature, modifier identity, modifier volume and sample matrix on the SCF extraction of several phenolic compounds were investigated. Solvent capacity was found to increase with pressure and with the amount of alcohol used as the modifier. Such variation in solvent capacity could be used to design a selective separation process where individual phenolic compounds, or groups of them, could be obtained.

SFE has been widely used in the extraction and recovery of high-value compounds, since the extract is obtained at relatively low temperature without any trace of organic solvent. The essential oil of aromatic herbs has usually been isolated by hydrodistillation, steam distillation or solvent extraction. The disadvantages of all these techniques are: low yield, losses of volatile compounds, long extraction times, toxic solvent residues and degradation of unsaturated compounds giving undesirable off-flavour compounds due to heat. Although hydrodistillation is a simple process characterized by low investment cost, loss of highly volatile and hydrosoluble substances is undeniable. Some of the differential features of SFE relative to other extraction techniques are summarized in Fig. 3.3.

SFE allows a continuous modification of solvent power and selectivity by changing the solvent density. Nevertheless, a simple SFE process consisting of SC-CO$_2$ extraction and a one-stage subcritical separation in many cases does not allow a selective extraction because of the simultaneous extraction of many unwanted compounds. This situation is typical of SC-CO$_2$ extraction of essential oils from herbaceous materials, in which even when the process is conducted under conditions that produce the optimum oil composition, cuticular waxes are co-extracted because of their lipophilic character and

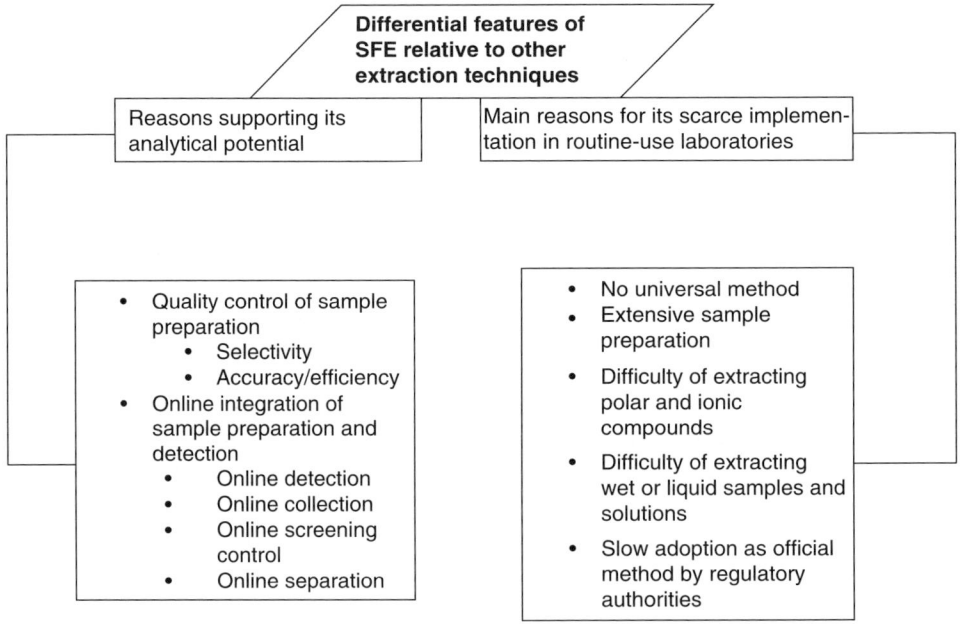

Fig. 3.3. Differential features of SFE relative to other extraction techniques (Zougagh *et al.*, 2004).

their localization on the leaf surface. SFE followed by fractional separation of the extract in multiple-stage separators overcomes these limitations and produces high-quality essential oils.

In recent years, SFE has received a great deal of attention as the full potential of this technology has begun to emerge. Nowadays, SFE has become an acceptable extraction technique and is used extensively in many areas. The major advantages of SFE in natural product studies are summarized as follows (Lang and Wai, 2001):

1. With comparable or better recoveries, the extraction time could be reduced from hours or days in a liquid–solid extraction to a few tens of minutes in SFE.
2. In SFE, quantitative or complete extraction could be ensured because the fresh fluid is continuously forced to flow through the samples.
3. The tunable solvation power of SCF is particularly useful for the extraction of complex samples from plant materials. Selective extraction of vindoline component from among more than 100 alkaloid compounds from the leaves of *Catharanthus roseus* is an example of selective extraction.
4. Sample concentration time can be eliminated as it is usually time-consuming and often results in loss of volatile components as solute dissolved in SCF can be separated easily by depressurization.
5. SFE can be an ideal technique for the study of thermolabile compounds as it is usually performed at low temperature. Sometimes, it could lead to the discovery of new natural compounds. Undesirable reactions such as hydrolysis, oxidation, degradation and rearrangement could be prevented effectively, for example in the extraction of ginger using SFE.
6. As compared to liquid–solid extraction, SFE can be carried out with smaller sample sizes. From only about 1.5 g of fresh plant samples, more than 100 volatile and semi-volatile compounds could be extracted and detected by gas chromatography–mass spectrometry (GC-MS), of which more than 80 compounds were in sufficient quantity for accurate quantifications.
7. SFE uses no or significantly less volume of environmentally hazardous solvents.
8. SFE may allow direct coupling with a chromatographic method, which can be a useful means to extract and directly quantify highly volatile compounds.
9. In large-scale SFE processes, SCFs, usually carbon dioxide, can be recycled or reused; thus, waste generation could be minimized.
10. SFE can be applied to systems of different scales: analytical scale (less than a gram to a few grams of samples), preparative scale (several hundred grams of samples), pilot plant scale (kilograms of samples) and large industrial scale (tonnes of raw materials).

Another distinguished advantage of SFE over conventional methods, in addition to the above-mentioned advantages, is that SFE can provide more information pertaining to the extraction processes and mechanisms. Such information could be used to assess quantitatively or evaluate the extraction efficiency and then for process optimization accordingly.

SFE provides the scope for online fractionation by manipulating the extraction conditions with a view to improving specific extractions. It is also possible to separate the extracted compounds into a group by adjusting operational parameters such as the type and proportion of liquid modifier or by altering the pressure/temperature of the SCF (Zougagh *et al.*, 2004). Palma *et al.* (1999) used a two-step separation technique for the investigation of active phenolic compounds in grape seeds. Fractions containing strong antioxidant activity consisting of fatty acids, aliphatic aldehydes and sterols were obtained using SC-CO$_2$. In the second step, SC-CO$_2$ modified with ethanol (20%) was used in fractions enriched in epicatechin and gallic acid. Sargenti and Lancas (1997) used a stepwise process SFE of lemon grass. First, the extraction was performed with 10 and 30% hexane, followed by another step with 10 and 20% acetone, and finally with 10% methanol. The extraction result with 10% hexane-modified CO$_2$ was similar to that obtained by steam extraction, while the result with 30% hexane was similar to that obtained with Soxhlet extraction in hexane. Acetone-modified CO$_2$ could extract additional compounds, and 10% methanol-modified CO$_2$ was not selective. SFE technique lacks a universal method that can be applied for all analytes and matrices. Although CO$_2$ is an excellent solvent for non-polar analytes, its most frequent limitation as an analytical extraction solvent is that its polarity is too low to obtain efficient extraction. Although the polarity of SC-CO$_2$ can be increased by adding modifier, selectivity is reduced because more impure compounds may be co-extracted with the target analyte.

3.6   SFE Process Upscaling

For scale-up and process design of the SFE extraction process, model description is an effective method. In this context, the following approaches were proposed (Roy *et al.*, 1994; Reverchon, 1997; Izadifar and Abdolahi, 2006; Han *et al.*, 2009):

1. Empirical models.
2. Single-sphere models.
3. Differential mass balance integration models.
4. Artificial network models.

The extraction process at bench scale was described by Sovova's extended Lack's model (Sovova, 1994). It is based on differential mass balance integration models (Papamichali *et al.*, 2000; Kiriamiti *et al.*, 2002). Both for bench and pilot plant scale, models fit only when calculations are reliable for the extraction process (Perrut *et al.*, 1997).

3.7 Supercritical Fluid Chromatography (SFC)

In the 1960s, Kelsper and colleagues proposed the use of supercritical carbon dioxide for eluting a chromatographic column and developed the first supercritical fluid chromatography (SFC) equipment. SFC primarily uses supercritical carbon dioxide as an eluent or mobile phase (Kelsper *et al.*, 1962). The physical characteristics exhibited by SCF include a diffusion coefficient of dissolved molecules that is 100 times greater than it is in liquid and lower viscosity. Because of the rapid mass transfer inside the column as compared to high-speed liquid chromatography using liquid as the mobile phase, separation is rapid without any drop in separation efficiency. Further, with the lower viscosity and higher diffusivity of supercritical carbon dioxide, higher mobile-phase velocity can be used in the column, leading to a higher process throughput as compared to liquid chromatography. The eluent leaving the column is then decompressed below its critical pressure and the supercritical solvent is transformed into a gas phase. When a modifier is used, the gaseous CO_2 is removed and the product is recovered in the liquid co-solvent.

As carbon dioxide is used as a medium, gasification will occur at a constant temperature from the separated and fractionated sample. This cuts the solvent cost and offers high-throughput analysis and preparative isolation due to rapid separation over a short period. For gradient elution, three parameters, namely pressure, temperature and modifier volume, can be varied in SFC.

Generally, all HPLC stationary phases can be used for SFC and are compatible with both HPLC and gas chromatography detectors.

SFC is recognized as a clean, green and efficient tool for analysis and purification. With the recent advances and accessibility of instrumentation, improved column hardware and wide variety of surface chemistries available, SFC has been finding an ever-increasing range of applications in many industries, such as: (i) pharmaceutical; (ii) nutraceutical; (iii) petrochemical; (iv) natural products; (v) food and beverages; and (iv) environmental.

HPLC is still the method of choice for chromatographic purifications and also remains the method of choice for reversed-phase separations because of the non-polar nature of carbon dioxide. It is often said that SFC is three to five times faster than HPLC. Much shorter retention time leads to increased productivity for SFC. Under supercritical conditions, the chromatographic column is equilibrated in a few minutes instead of hours. Additionally, longer columns, packed with smaller particles can be used in contrast to HPLC. Retention in SFC separations is more easily adjustable than in HPLC. For the screening of optimal operating conditions, pressure and co-solvent additions are the most frequently evaluated parameters. As the temperature effect is stronger in SFC as compared to HPLC, it can be used to fine tune selectivity. SFC is more flexible with respect to a suitable detector. All HPLC detectors equipped with high-pressure cells, but also many common GC detectors, can be used either online or offline to analyse column effluent composition (Majewski *et al.*, 2005).

3.8 Conclusion

SFE technology requires relatively high capital investment; however, the cost of preparing SFE extracts is competitive with traditional technologies. Lack of a universal method that works for all analytes and matrices is one of the major limitations of SFE technology. Slow adoption as an official technique by a regulatory authority further restricts the demand for this technology.

References

Abbas, K.A., Mohamed, A., Abdulamir, A.S. and Abas, H.A. (2008) A review on super critical fluid extraction as new analytical method. *American Journal of Biochemistry and Biotechnology* 4, 345–353.

Andersen, M.R., Swanson, J.J., Porter, N.L. and Richter, B.E. (1989) Supercritical fluid extraction as a sample introduction method for chromatography. *Journal of Chromatographic Sciences* 27, 371–377.

Ashraf-Khorassani, M., Giganian, S. and Yamini, Y. (1995) Effect of pressure, temperature, modifier, modifier concentration, and sample matrix on the supercritical fluid extraction efficiency of different phenolic compounds. *Journal of Chromatographic Sciences* 33, 658–662.

Beckman, E.J. (2004) Supercritical and near-critical CO_2 in green chemical synthesis and processing. *Journal of Supercritical Fluids* 28, 121–191.

Berger, T.A. (1989) Effects of temperature and density on retention in capillary supercritical-fluid chromatography. *Journal of Chromatography* 478, 311–324.

Bezerra, M.A., Santelli, R.E., Oliveira, E.P., Villar, L.S. and Escaleira, L.A. (2008) Response surface methodology (RSM) as a tool for optimization in analytical chemistry. *Talanta* 76, 965–977.

Bicchi, C., Rubiolo, P., Frattini, C., Sandra, P. and David, F. (1991) Off-line supercritical fluid extraction and capillary gas chromatography of pyrrolizidine alkaloids in *Senecio* species. *Journal of Natural Products* 54, 941–945.

Bjorklund, E., Jaremo, M., Mathiasson, L., Jansson, J.A. and Karlsson, L. (1998) Illustration of important mechanisms controlling mass transfer in supercritical fluid extraction. *Analytica Chimica Acta* 368, 117–128.

Bravi, E., Perretti, G., Monanari, L., Favati, F. and Fantozzi, P. (2007) Supercritical fluid extraction for quality control in beer industry. *Journal of Supercritical Fluids* 42, 342–346.

Capriel, P., Haisch, A. and Khan, S.U. (1986) Supercritical methanol: an efficacious technique for the extraction of bound pesticide residues from soil and plant samples. *Journal of Agricultural and Food Chemistry* 34, 70–73.

Casas, L., Mantell, C., Rodroguez, M., Torres, A., Macias, F.A., *et al.* (2007) Effect of the addition of cosolvent on the supercritical extraction of bioactive compounds from *Helianthus annus*. *Journal of Super Critical Fluids* 41, 43–49.

Casas, L., Mantell, C., Rodroguez, M., Torres, A., Macias, F.A., *et al.* (2008) Super critical extraction of bioactive compounds from sunflower leaves with carbon dioxide and water on a pilot plant scale. *Journal of Super Critical Fluids* 45, 37–42.

Casas, L., Mantell, C., Rodríguez, M., López, E. and Ossa, E.M. (2009) Industrial design of multifunctional supercritical extraction plant for agro-food raw materials. *Chemical Engineering Transactions* 17, 1585–1590.

Catchpole, O.J., Grey, J.B., Perry, N.B., Burgress, E.J., Redmond, W.A., *et al.* (2003) Extraction of chili, black pepper and ginger with near critical CO_2, propane and dimethyl ether: analysis of extracts by quantitative nuclear magnetic resonance. *Journal of Agricultural and Food Chemistry* 51, 4853–4860.

Chen, Q., Yao, S., Huang, X., Luo, J., Wang, J., *et al.* (2009) Supercritical fluid extraction of *Coriandrum sativum* and subsequent separation of isocoumarins by high speed counter current chromatography. *Food Chemistry* 117, 504–508.

Chun, M.K., Shin, H.W. and Lee, H. (1994) Supercritical fluid extraction of taxol and baccatin from needle of taxus cuspidate. *Biotechnology Techniques* 8, 547–550.

Daneshfar, A., Barzegar, M., Ashraf-Khorassani, M. and Levy, J.L. (1995) Supercritical fluid extraction of phenoxy acids from water. *Journal of High Resolution Chromatography* 18, 446–448.

Diaz, M.S. and Brignole, E.A. (2009) Modeling and optimization of supercritical fluid processes. *Journal of Super Critical Fluids* 47, 611–618.

Engelhaqrdt, H. and Haas, P. (1993) Possibilities and limitations of SFE in the extraction of Aflatoxin B1 from food matrices. *Journal of Chromatographic Sciences* 31, 13–19.

Espinosa, S., Diaz, M.S. and Brignole, E.A. (2005) Process optimization for super critical concentration of orange peel oil. *Latin American Applied Research* 35, 321–326.

Esquivel, M., Ribeiro, M. and Bernardo-Gil, M. (1999) Supercritical extraction of savory oil: study of antioxidant activity and extract characterization. *Journal of Supercritical Fluids* 14, 129–138.

Filho, G.L., De Rosso, V.V., Meireles, M.A.A., Rosa, P.T.V., Oliveira, A.L., *et al.* (2008) Supercritical fluid extraction of carotenoids from Pitango fruits (*Eugenia uniflora* L). *Journal of Supercritical Fluids* 46, 33–39.

Furton, K.G. and Rein, J. (1991) Effect of micro extractor cell geometry on supercritical fluid extraction recoveries and correlations with supercritical fluid chromatographic data. *Analytica Chimica Acta* 248, 263–270.

Gorbaty, Y.E. and Bondarenko, G.V. (1998) The physical state of supercritical fluids. *Journal of Super Critical Fluids* 14, 1–8.

Guçlu-Ustundag, O. and Temelli, F. (2004) Correlating the solubility behaviour of minor lipid components in supercritical carbon dioxide.*The Journal of Supercritical Fluids* 31, 235–253.

Han, X., Cheng, L., Zhang, R. and Bi, J. (2009) Extraction of safflower seed oil by supercritical CO_2. *Journal of Food Engineering* 92, 370–376.

Hawthorne, S.B., Riekkola, M., Serenius, K., Holm, Y., Hiltunen, K., *et al.* (1993) Comparison of hydrodistillation and supercritical fluid extraction for the determination of essential oils in aromatic plants. *Journal of Chromatography A* 634, 297–308.

Hutz, A., Schmidt, F.P., Leyendecker, D. and Kelper, E. (1990) Effects of temperature, pressure, and density on the chromatographic behavior of supercritical carbon dioxide. *Journal of Supercritical Fluids* 3, 1–7.

Izadifar, M. and Abdolahi, F. (2006) Comparison between neural network and mathematical modeling of supercritical CO_2 extraction of black pepper essential oil. *Journal of Supercritical Fluids* 38, 37–43.

Jacobson, G.B., Moulder, R., Lu, L., Bergstro, M., Markides, K.E., *et al.* (1997) Super critical fluid extraction of ^{11}C-labeled metabolites in tissues using super critical ammonia. *Analytical Chemistry* 69, 275–280.

Jeong, M.L. and Chesney, D.J. (1999) Investigation of modifier effects in super critical CO_2 extraction from various solid matrices. *Journal of Supercritical Fluids* 16, 33–42.

John, S., Chun, Y., Sophia, J.X., Yueming, J., Ying, M., *et al.* (2009) Effects of modifiers on the the profile of lycopene extracted from tomato skins by super critical carbon dioxide. *Journal of Food Engineering* 93, 431–436.

Joyces, P.C. and Thies, M.C. (1997) Separation of fischer-tropsch wax from catalyst by super critical extraction. Quarterly Report, 1 October–31 December. US Department of Energy Office of Fossil Energy (DE-FG22-94PC94219-13).

Kiriamiti, H.K., Rascol, E., Marty, A. and Condoret, J.S. (2002) Extraction rates of oil from high oleic sunflower seeds with supercritical carbon dioxide. *Chemical Engineering and Processing* 41, 711–718.

Kelsper, E., Corvin, A.H. and Turner, D.A. (1962) High pressure gas chromatography above critical temperatures. *Journal of Organic Chemistry* 27, 700–701.

Krichnavaruk, S., Shotipurk, A., Goto, M. and Pavasant, P. (2008) Supercritical carbon dioxide extraction of astaxanthin from *Haematococcus pluvialis* with vegetable oils as co solvent. *Bioresource Technology* 99, 5556–5560.

Kumoro, A.C. and Hasan, M. (2007) Supercritical carbon dioxide extraction of andrographolide from *Andrographis paniculata*: effect of the solvent flow rate, pressure and temperature. *Chinese Journal of Chemical Engineering* 15, 877–883.

Lang, Q.Y. and Wai, C.M. (2001) Supercritical fluid extraction in herbal and natural product studies – a practical review. *Talanta* 53, 771–782.

Langenfeld, J.J., Hawthrone, S.B., Miller, D.J. and Pawliszyn, J. (1993) Effects of temperature and pressure on supercritical fluid extraction efficiencies of polycyclic aromatic hydrocarbons and polychlorinated biphenyls. *Analytical Chemistry* 65, 338–334.

Leal, P.F., Maia, N.B., Carmello, Q.A.C., Catharino, R.R., Eberlin, M.N., *et al.* (2008) Sweet basil (*Ocimum basilicum*) extracts obtained by super critical fluids extraction (SFE): global yields, chemical composition, antioxidant activity and estimation of the cost of manufacturing. *Food and Bioprocess Technology: An International Journal* 1, 326–328.

Lehoty, S.J. (1997) Supercritical fluid extraction of pesticides in foods. *Journal of Chromatography A* 785, 289–312.

Liang, S. and Tilotta, D.C. (1998) Extraction of petroleum hydrocarbons from soil using super critical argon. *Analytical Chemistry* 70, 616–622.

Liu, J., Lin, S.Y., Wang, Z.Z., Wang, C.N., Wang, E.L., *et al.* (2011) Supercritical fluid extraction of flavonoids from *Maydis stigma* and its nitrite scavenging ability. *Food Bioproduct Processing* 89, 333–339.

Ma, Q., Xu, X., Gao, Y., Wang, Q. and Zhao, J. (2008) Optimization of super critical carbon dioxide extraction of lutein from Marigold (*Tagetes erect* L.) with soybean oil as cosolvent. *International Journal of Food Science and Technology* 43, 1763–1769.

Machmudah, S., Kawahito, Y., Sasaki, M. and Goto, M. (2007) Super critical CO_2 extraction of rosehip seed oil fatty acids composition and process optimization. *Journal of Supercritical Fluids* 41, 421–428.

Machmudah, S., Kawahito, Y., Sasaki, M. and Goto, M. (2008) Process optimization and extraction rate analysis of carotenoids extraction from rosehip fruit using super critical CO_2. *Journal of Supercritical Fluids* 44, 308–314.

Majewski, W., Valery, E. and Ludemann-Hombourger, O. (2005) Principle and applications of supercritical fluid chromatography. *Journal of Liquid Chromatography and Related Technologies* 28, 1233–1252.

Modey, W.K., Mulholland, D.A. and Raynor, M.W. (1996) Analytical supercritical fluid extraction of natural products. *Phytochemical Analysis* 7, 1–15.

Mohamed, R.M., Saldana, M.D.A. and Mazzafera, P. (2002) Extraction of caffeine, theobromine and cocoa butter using super critical CO_2 and ethane. *Industrial Engineering and Chemical Research* 41, 6751–6758.

Mostafa, K., Yadollah, Y., Fatermeh, S. and Naader, B. (2004) Comparison of essential oil composition of *Carum copticum* obtained by supercritical carbon dioxide and hydro-distillation methods. *Food Chemistry* 86, 587–591.

Murga, R., Ruiz, R., Beltran, S. and Cabezas, J.L. (2000) Extraction of natural complex phenols and tannins from grape seeds by using supercritical mixtures of carbon dioxide and alcohol. *Journal of Agricultural and Food Chemistry* 48, 3408–3412.

Mushtaq, M.Y., Choi, Y.H., Verpoorte, R. and Wilson, E.G. (2014) Extraction for metabolomics: access to the metabolome. *Phytochemical Analysis* 25, 291–306.

Nemoto, S., Sasaki, K., Toyoda, M. and Saito, Y. (1997) Effect of extraction conditions and modifiers on supercritical fluid extraction of 88 pesticides. *Journal of Chromatographic Science* 35, 467–477.

Oman, M., Skerget, M. and Knez, Z. (2013) Application of supercritical fluid for the separation of nutraceuticals and other phytochemicals from plant material. *Macedonian Journal of Chemistry and Chemical Engineering* 32, 183 226.

Onuska, F.I. and Terry, K.A. (1989) Supercritical fluid extraction of PCBs in tandem with high resolution gas chromatography in environmental analysis. *Journal of High Resolution Chromatography* 12, 527–531.

Oostdyk, T.S., Grob, R.L., Synder, M.E. and McNally, J. (1993) Optimization of the supercritical fluid extraction of primary aromatic amines. *Journal of Chromatographic Sciences* 31, 177–182.

Palma, M., Talor, L.T., Varela, R.M., Cutler, S.J. and Cutler, H.G. (1999) Fractional extraction of compounds from grape seeds by supercritical fluid extraction and analysis for antimicrobial and agrochemical activities. *Journal of Agricultural and Food Chemistry* 47, 5044–5048.

Papamichali, I., Louli, V. and Magoulas, K. (2000) Super critical fluid extraction of celery seed oil. *Journal of Supercritical Fluids* 18, 213–226.

Pasquali, I., Bettine, R. and Giordano, F. (2006) Solid state chemistry and particle engineering with supercritical fluids in pharmaceutics. *European Journal of Pharmaceutical Sciences* 27, 299–310.

Perakis, C., Louli, V. and Magoulas, K. (2005) Supercritical fluid extraction of black pepper oil. *Journal of Food Engineering* 71, 386–393.

Pereira, C. and Meireles, M.A. (2010) Super critical fluid extraction of bioactive compounds: fundamentals, applications and economic perspectives. *Food Bioprocess Technology* 3, 340–372.

Perrut, M., Clavier, J.Y., Poletto, M. and Reverchon, E. (1997) Mathematical modeling of sunflower seed extraction by super critical CO_2. *Industrial Engineering Chemistry Research* 36, 430–435.

Phelps, C.L., Smart, N.G. and Wai, C.M. (1996) Past, present and possible future applications of super critical extraction technology. *Journal of Chemical Education* 73, 1163.

Pourmortazavi, S.M. and Hajimirsadeghi, S.S. (2007) Supercritical fluid extraction in plant essential and volatile oil analysis. *Journal of Chromatography A* 1163, 2–24.

Quispe-Condori, S., Foglio, M.A., Rosa, P.T.V. and Meireles, M.A.A. (2008) Obtaining β-caryophyllene from *Cordia Verbencea de Condoule* by supercritical fluid extraction *Journal of Supercritical Fluids.* 46, 27–32.

Reverchon, E. (1997) Supercritical fluid extraction and fractionation of essential oils and related products. *Journal of Supercritical Fluids* 10, 1–37.

Reverchon, E. and De Marco, I. (2006) Supecritical fluid extraction and fractionation of natural matter. *Journal of Supercritical Fluids* 38, 146–166.

Roy, B.C., Goto, M., Hirose, T., Navaro, O. and Hortacsu, O. (1994) Extraction rates of oil from tomato seeds with super critical carbon dioxide. *Journal of Chemical Engineering of Japan* 27, 768–772.

Sanal, I.S., Bayraktar, E., Mehmetoglu, U. and Calimlt, A. (2005) Determination of optimum conditions of SC-(CO_2+ ethanol) extraction of β-carotene from apricot pomace using response surface methodology. *The Journal of Supercritical Fluids* 34, 331–338.

Sargenti, S.R. and Lancas, F.M. (1997) Super critical fluid extraction of *Cymbopogon citrates* (DC) Staff. *Chromatographia* 46, 285–290.

Sauceau, M., Letourneau, J.J., Freiss, B., Richon, D. and Fages, J. (2004) Solubility of eflucimibe in supercritical carbon dioxide with or without a co-solvent. *Journal of Supercritical Fluids* 31, 133–140.

Shende, R.V. and Lombardo, S.J. (2002) Supercritical extraction with carbon dioxide and ethylene of poly(vinyl butrayl) and dioctyl phthalate from multilayer ceramic capacitors. *Journal of Supercritical Fluids* 23, 885–888.

Shi, J., Yi, C., Xue, S.J., Jiang, Y., Ma, Y., *et al.* (2009) Effects of modifiers on the profile of lycopene extracted from tomato skins by supercritical carbon dioxide. *Journal of Food Engineering* 93, 431–436.

Smith, R.M. and Burford, M.D. (1992) Optimization of supercritical fluid extraction of volatile constituents from a model plant matrix. *Journal of Chromatography A* 600, 175–181.

Solati, Z., Baharin, B.S. and Baghreri, H. (2012) Supercritical carbon dioxide (SC-CO$_2$) extraction of *Nigella sativa* L. oil using full factorial design. *Industrial Crops and Products* 36, 519–523.

Sovilj, M.N., Nikolovski, B.G. and Spasojevic, M.D. (2011) Critical review of supercritical fluid extraction of selected spice plant materials. *Macedonian Journal of Chemistry and Chemical Engineering* 30, 197–220.

Sovova, H. (1994) Rate of vegetable oil extraction with supercritical carbon dioxide – I modeling of extraction curves. *Chemical Engineering Science* 49, 409–414.

Sovova, H. (2005) Mathematical model for supercritical fluid extraction of natural products and extraction curve evaluation. *Journal of Supercritical Fluids* 33, 35–52.

Stahl, E. and Glatz, A. (1994) Extraction of natural substances with supercritical gases. 10. Communication: qualitative and quantitative determination of solubilities of steroids in super critical carbon dioxide. *Fette Seifen Anstrichmittel* 86, 346–348.

Subramaniam, B., Rajewski, R.A. and Snavely, K. (1993) Pharmaceutical processing with supercritical carbon dioxide. *Journal of Pharmaceutical Science* 86, 885–890.

Sun, M. and Temeli, F. (2006) Super critical carbon dioxide extraction of carotenoids from carrot using canola oil as a continuous co-solvent. *Journal of Supercritical Fluids* 37, 397–408.

Tan, C.H., Ghazali, H.M., Kuntom, A., Tan, C.P. and Ariffin, A.A. (2009) Extraction and physicochemical properties of low free fatty acid crude palm oil. *Food Chemistry* 113, 645–650.

Taylor, S.L., King, J.W. and Geer, J.J. (1993) Analytical-scale supercritical fluid extraction of aflatoxin B1 from field-inoculated corn. *Journal of Agricultural and Food Chemistry* 41, 910–913.

Tena, M.T., Valacel, M., Hidalgo, P.J. and Ubera, J.L. (1997) Supercritical fluid extraction of natural antioxidants from rosemary: comparison with liquid solvent sonication. *Analytical Chemistry* 69, 521–526.

Teruel, M.L.A., Gontier, E., Bienaime, C., Saucedo, J.E.N. and Barbotin, J.N. (1997) Response surface analysis of chlortetracycline and tetracycline production with K-carrageenan immobilized *Streptomyces aureofaciens*. *Enzyme and Microbial Technology* 21, 314–320.

Turner, C., King, J.W. and Mathiasson, L. (2001) Super critical extraction and chromatography for fat soluble vitamin analysis. *Journal of Chromatography A* 936, 215–237.

Vasapollo, G., Longo, L., Rescio, L. and Ciurlia, L. (2004) Innovative supercritical carbon dioxide extraction of lycopene from tomato in presence of vegetable oil as a cosolvent. *Journal of Supercritical Fluids* 29, 87–96.

Vazquez, L., Torres, C.F., Fornari, T., Senorans, F.J. and Reglero, G. (2007) Recovery of squalene from vegetable oil sources using countercurrent supercritical carbon dioxide extraction. *Journal of Supercritical Fluids* 40, 59–66.

Wakte, P.S., Sachin, B.S., Patil, A.A., Mohato, D.M., Band, T.H., *et al.* (2011) Optimization of microwave, ultrasonic and super critical carbon dioxide assisted extraction techniques for curcumin from *Curcuma longa*. *Separation and Purification Technology* 79, 50–55.

Walsh, J.M., Ikonomou, G.D. and Donohue, M.D. (1987) Supercritical x behavior: the entrainer effect. *Fluid Phase Equilibria* 33, 295–314.

Wang, H.C., Chen, C.R. and Chang, C.J. (2001) Carbon dioxide extraction of ginseng root hair oil and ginsenosides. *Food Chemistry* 72, 505–509.

Wang, H., Liu, Y., Wei, S. and Yan, Z. (2012) Application of response surface methodology to optimize super critical carbon dioxide extraction of essential oil from *Cyperus rotundus* Linn. *Food Chemistry* 132, 582–587.

Weathers, R.M., Beckholt, D.A., Lavella, A.L. and Danielson, N.D. (1999) Comparison of acetals as *in situ* modifiers for the supercritical fluid extraction of beta-carotene from paprika with carbon dioxide. *Journal of Liquid Chromatography and Related Technologies* 22, 241–252.

Wood, J.A., Bernards, M.A., Wan, W.K. and Charpentier, P.A. (2006) Extraction of ginsenosides from North American ginseng using modified super critical carbon dioxide. *Journal of Supercritical Fluids* 39, 40–17.

Wright, B.W., Wright, C.W., Gale, R.W. and Smith, R.D. (1989) Analytical supercritical fluid extraction of adsorbent materials. *Analytical Chemistry* 59, 38–44.

Yonker, C.R., Frye, S.L., Kalkwarf, D.R. and Smith, R.D. (1986) Characterization of supercritical fluids solvents using solvatochromic shifts. *Journal of Physical Chemistry* 90, 3022–3026.

Zizovic, I., Stamenic, M., Ivanovic, J., Orlovic, A., Ristic, M., *et al.* (2007) Supercritical carbon dioxide extraction of sesquiterpenes from Valerian root. *Journal of Supercritical Fluids* 43, 249–258.

Zosel, K. (1969) German Patent 1, 493, 190.

Zougagh, M., Valcarcel, M. and Rios, A. (2004) Supercritical fluid extraction: a critical review of its analytical usefulness. *Trends in Analytical Chemistry* 23, 399–340.

4

Chromatographic Techniques: High-performance Thin-layer Chromatography and High-speed Countercurrent Chromatography

4.1 Introduction

Before the establishment of instrumental chromatographic methods such as gas chromatography and high-performance liquid chromatography, thin-layer chromatography (TLC) was the most common method used for the analysis of natural products. Still today, TLC is the most frequently used semi-quantitative method for the preliminary identification of constituents. Here, separation is achieved on the basis of partition or a combination of partition and adsorption, depending on the composition of the stationary phase and the mobile phase (Stahl, 1969). High-performance thin-layer chromatography (HPTLC) is an automated form of TLC with better and more advanced separation capacity and detection limit. HPTLC is also known as planar chromatography or flat-bed chromatography (Srivastava, 2011; Sethi, 2013). Nowadays, HPTLC is used as an alternative analytical technique to gas chromatography (GC) and high-performance liquid chromatography (HPLC). Further, its hyphenation with other advanced techniques such as spectroscopic techniques to provide HPTLC-mass spectrometry (HPTLC-MS) and HPTLC-Fourier transform infrared spectroscopy (HPTLC-FTIR), scanning diode laser, etc., has made HPTLC a powerful analytical tool in the area of natural product research. Quantification at a very low level (nanogram), better accuracy and sensitivity, and minimization of handling error due to automation may be listed as some of the advantages of HPTLC. However, technical expertise is required for proper utilization of this rapid analytical tool. Applications of HPTLC include: quality control, metabolic studies, bioassays, etc. HPTLC has some unique advantages over HPLC. Sample treatment is simple because of single use of the layer, analysis time is shorter and multiple samples can be run with standards in a single plate. Furthermore, low volume of solvent is consumed. Simultaneous chromatography of samples and

  © Satyanshu Kumar 2016. *Analytical Techniques for Natural Product Research* (S. Kumar)

standards under identical conditions on the same layer leads to results with accuracy and precision.

Countercurrent chromatography (CCC) is an automated version of liquid–liquid extraction, comparable to the repeated partitioning of an analyte between the two immiscible phases by vigorous mixing in a separatory funnel. Modern CCC began in the 1970s with the development of droplet countercurrent chromatography (DCCC). DCCC has been applied successfully to the isolation of natural compounds such as saponins, alkaloids, peptides and glycosides (Hostettmann, 1980; Winterhalter *et al.*, 1990). However, the DCCC technique also had some disadvantages. Solvent systems for DCCC were limited to those able to form droplets, a characteristic that was influenced by the flow rate, surface tension and viscosity. As only low flow rates could be applied, the separation time was long and leakage at the several hundred connections was also a common problem. Moreover, separation efficiency was also low because of the poor mixing of the two phases. Therefore, to improve mixing of the two phases, techniques known as rotation locular CCC (RLCCC) and gyration locular CCC (GLCCC) were developed. The real breakthrough of CCC came with the invention of the coil planet centrifuge, which was introduced by Ito (1981). Today, this technique is known as high-speed CCC (HSCCC) and is based on the principles of liquid–liquid partitioning chromatography.

4.2   Principles and Workings of HSCCC

Two immiscible liquid phases are mixed together by centrifugal force and pressure to form a two-phase system and then separated multiple times. It is at this time that the exchange of molecules between the two phases occurs. The separation of the solutes is achieved as a function of the specific partitioning coefficient of each solute between the mobile and stationary phases. The individual solutes are isolated based on the different partitioning coefficients of each compound in this two-phase system.

HSCCC separation takes place in a so-called multilayer coil that is made by wrapping inert Teflon tubing around a holder in multiple layers. Consequently, the technique is also known as multilayer coil CCC (MLCCC).

HSCCC is a type of support-free, all-liquid partition chromatography, which eliminates the irreversible adsorption of samples on the solid support that occurs in conventional chromatography (Ito, 1991), the stationary phase of which is retained in the separation column by gravity and the centrifugal field (Ito, 1991; Wang and Liu, 2007). HSCCC avoids the disadvantages arising from the interaction of samples with the solid support, such as the absorption and denaturation of target products. It also has the unique features of high recovery and high efficiency.

An HSCCC system is similar to an HPLC system and consists of a mobile-phase reservoir, pump, injection valve, a column, a fraction collector and a data processor. The liquid stationary phase is held in an inert, coiled tubular column by a centrifugal force field when the immiscible mobile phases pass through (Berthod *et al.*, 2003).

Separation in HSCCC is based purely on the partition of the solutes between the two liquid phases as no solid support matrix is employed to retain the stationary phase. Solutes have access to the whole volume of the stationary phase rather than the interface between the mobile phase and the surface of the solid stationary phase, leading to high efficiency of extraction. Either phase of the two-phase solvent system can be used as the mobile phase. The phase role can even be changed during a run to elute the entire injected sample from the column.

4.3 Solvent System of HSCCC

An appropriate solvent system plays an important role in separation by HSCCC. The solvent system for HSCCC is selected based on the differences in the partition coefficient (K) of two or more components (Sun et al., 2009). The selection of the two-phase solvent system for the target compounds is an important step in HSCCC, as successful separation by this technique depends on the selection of a suitable two-phase solvent system that provides an ideal range of coefficients (K) for the target compounds. The composition of the solvent system is selected according to K values of the target compounds in the sample. The best isolation conditions are optimized with regard to best possible separation and run time considerations. While selecting a two-phase solvent, the following are taken into consideration (Ito, 1988; Oka et al., 2000, 2001):

1. Retention of the stationary phase should be satisfactory.
2. The settling time of the solvent system should be short (<30 s).
3. The partition coefficient (K) of the target compound should be 1.
4. The separation factor (α) between two compounds should be greater than 1.5.

After investigating the effects of the two-phase solvent system, flow rate and revolution speed of the separation column are the other factors for consideration. The partition coefficient is the most important parameter in solvent system selection. The partition coefficient reflects the solute distribution between the two mutually equilibrated solvent phases. It should be close to 1 to get an efficient separation and suitable run time. If it is much smaller than 1, the solutes will be eluted close to each other near the solvent front, resulting in the loss of peak resolution. If the K value is much greater than 1, the solutes will be eluted in excessively broad peaks (band broadening), which may lead to extended elution time (Chen et al., 2003).

The K values of the target compounds in different solvent systems are determined by HPLC as elucidated by the following example. A suitable amount of crude sample is dissolved in the upper phase of the equilibrated two-phase solvent system. Thereafter, the contents of the solution are determined by HPLC and the peak areas recorded as A. Then an equal volume of organic phase is added to the solution and mixed thoroughly. After equilibration is established, the lower phase contents are also determined by

HPLC again and the peak areas recorded as B. The partition coefficient (K) is obtained by the following equation $K = A/B$. Determination of the partition coefficient for a two-phase solvent system composed of chloroform–methanol–water (4:3.5:2, v/v) used for the isolation and purification of three flavonoids from the Chinese medicinal plant, *Epimedium koreanum*, was described by Liu *et al.* (2005b). A suitable amount of crude extract was dissolved in the lower phase and the content of the solution was determined by HPLC (peak area A_1). Then an equal volume of the upper phase was added to the solution and mixed thoroughly. After partition equilibration was reached, the lower-phase solution was determined by HPLC again and the peak area was recorded as A_2. The K values were calculated according to the following equation: $K = (A_1 - A_2)/A_2$ (Sun *et al.*, 2009; Table 4.1).

Optimal solvent should satisfy two requirements. First, the value of the partition coefficient (K) of the solvent system should be in the range of 0.5–2.0 (Liu *et al.*, 2005b; Wu *et al.*, 2005; Tables 4.1 and 4.2). Second, the higher the retention of the stationary phase, the better the peak resolution, because the retention of the stationary phase is accomplished by a combination of coiled column configuration and the planetary motion of the column holder (Ito, 2005).

The followings are examples of some two-phase solvent systems:

- light petroleum (boiling point 60–90°C)–methanol–water (5:4:1, v/v);
- chloroform–methanol–water (5:4:1, v/v);
- *n*-hexane–ethyl acetate–methanol–water (3:7:5:5, v/v); and
- *n*-hexane–ethyl acetate–methanol–water (6:4:5:5, v/v).

Table 4.1 Partition coefficient (K) values of paeonol in different solvent systems. (From Sun *et al.*, 2009.)

Solvent system	K
n-Butanol–methanol–water (4:1:5)	32.71
Ethyl acetate–methanol–water (3:2:5)	26.64
n-Hexane–methanol–water (4:1:5)	3.85
Petroleum ether–ethyl acetate–methanol–water (2:6:3:4)	0.96
Petroleum ether–ethyl acetate–methanol–water (5:5:5:6)	0.62
Petroleum ether–ethyl acetate–methanol–water (2:5:4:3)	1.24
Petroleum ether–ethyl acetate–methanol–water (2:6:3:5)	0.83

Table 4.2. K values of baicalin and wogonoside and other compounds in ethyl acetate–methanol–acetic acid (1%)–water. (From Wu *et al.*, 2005.)

Solvent system	K (a*)	K (b*)	K (baicalin)	K (c*)	K (d*)		K (wogonoside)
5:0.5:5	0.03	0.12	0.54	0.99	1.62	1.86	2.78
5:1:5	0.11	0.19	0.68	1.12	1.43	1.64	1.96
5:2:5	0.24	0.27	0.84	1.26	1.27	1.40	1.43

4.4   HSCCC Flow Condition

For development of HSCCC methods, the isocratic elution mode is mostly used, as any change in the composition of the mobile phase will change the composition of the stationary phase or volume. However, in some cases, HSCCC using an isocratic elution needs to be used repeatedly with the same or a different two-phase solvent system to obtain compounds of suitable purity.

4.5   Detection System of HSCCC

A UV detector is usually used for HSCCC. However, evaporative light scattering detectors (ELSDs) and mass spectrometric detectors are also used to monitor the HSCCC separation process. The sample may be dissolved in either phase or in a mixture of two phases. The injection volume is usually less than 5% of the total column capacity. The purity of the compound is a critical element in the whole process of obtaining the desired biologically active material needed for screening and for subsequent formation of the structure activity relationship (Leister *et al.*, 2003). Usually, each fraction after HSCCC purification is collected based on the UV response and then purity is determined by the HPLC or HPTLC technique. Online purity monitoring of the HSCCC fractions has been reported (Zhou *et al.*, 2006, 2007). The benefit of online purity monitoring is that it decreases the instrument idle time.

4.6   Applications of HSCCC

HSCCC is mainly a preparative purification technique. Crude extracts or semi-pure fractions can be chromatographed with sample loads ranging from milligrams to multigrams. The conventional methods for the purification of active compounds from natural products, such as recrystallization, column elution and preparative HPLC, etc., require tedious steps, resulting in low recoveries. HSCCC successfully eliminates irreversible adsorption or chemical reaction of samples with the solid support (Conway, 1990). In recent years, successful application of HSCCC has been reported for the purification of various kinds of active compounds such as alkaloids, flavonoids, coumarins, anthraquinones, phenolic acids, diterpenoids, saponins, etc. (Peng *et al.*, 2005, 2008; Sun *et al.*, 2009; Tang *et al.*, 2008; Zhu *et al.*, 2008; Hou *et al.*, 2009; Xin *et al.*, 2009; Yuan *et al.*, 2009). A brief description of application of HSCCC in phytochemistry is summarized in Table 4.3.

   With almost limitless possibilities for the selection of a two-phase solvent system, both aqueous and non-aqueous solvent systems, HSCCC has a unique advantage in the isolation and purification of natural products. It is a powerful technique for the isolation of bioactive compounds from plant materials, and also online purity monitoring can improve the

Table 4.3. Application of HSCCC in phytochemistry.

Serial number	Compound	Source	Reference
1	Alkaloids	*Evodia rutaecarpa*	Liu *et al.*, 2005a
2	Anthocyaninins	Red wine	Degenhardt *et al.*, 2000a,b
3	Anthraquinone	*Aloe vera*	Cao *et al.*, 2007
		Cassia tora	Zhu *et al.*, 2008
4	Carotenoids	*Gardenia jasminoides*	Degenhardt *et al.*, 2001
5	Catechins and proanthocyanidins	Black tea	Degenhardt *et al.*, 2000c
6	Chromone	*Radix saposhnikoviae*	Liu *et al.*, 2008
7	Coumarin	*Cnidium*	Wei *et al.*, 2004
8	Flavonoids	*Epimedium koreanum*	Liu *et al.*, 2005b
	Flavonol glycosides	*Alpinia katsumadai hyata*	Li *et al.*, 2007
		Oroxylum indicum	Chen *et al.*, 2005
		Epimedium brevicornum	Li and Chen, 2009
		Black tea, endive, shallots	Tang *et al.*, 2013
		Lophatherum gracile	Peng *et al.*, 2005
		Patrinia villosa	
9	Isoflavones	Soy flour	Degenhardt *et al.*, 2001
			Du *et al.*, 2001
10	Isocoumarins	*Coriandrum sativum*	Chen *et al.*, 2009
11	Lactones	*Aucklandia lappa* Decne	Li *et al.*, 2005
12	Lignans	Flaxseed	Degenhardt *et al.*, 2002
		Taraxacum mongolicum	Shi *et al.*, 2008
13	Phenolics	*Magnoliae officianalis*	Lu *et al.*, 2006
		Forsythia suspense	Li and Chen, 2005
		Smilax china	Yang *et al.*, 2008
14	Saponins	*Clematis mandshurica*	Shi *et al.*, 2007
		Platycodi radix	Kim and Ha, 2009
15	Xanthones	*Gentianella amarella*	Urbain *et al.*, 2008
		Garcinia mangostana	Destandau *et al.*, 2009
		Swertia mussotii	Jia *et al.*, 2012

efficiency of the overall purification process. HSCCC as an all-liquid chromatographic technique is operated under gentle conditions and allows the non-destructive isolation of labile natural compounds. Due to the absence of any solid stationary phase, adsorption losses are minimized; hence, a 100% sample recovery is guaranteed (Schwarz *et al.*, 2003). This technique has been applied successfully to the separation and isolation of many natural products (Zhou *et al.*, 2006, 2007; Lu *et al.*, 2007a,b; Wang and Liu, 2007; Zhao *et al.*, 2007). As an advanced separation technique, HSCCC has been widely used for the separation of active components in traditional Chinese herbs and other natural products in recent years. Successful application of HSCCC has also been reported for the separation and quantification of alkaloids (Liu *et al.*, 2005a), quinines (Du *et al.*, 2001), flavonoids (Chen *et al.*, 2005), coumarins (Wei *et al.*, 2004) and other natural products (Yao *et al.*, 2007).

One of the major factors restricting the use of HSCCC as an analytical tool is the speed at which separation may be conducted. A new small coil volume HSCCC as rapid as HPLC achieving high resolutions in minutes as opposed to hours was reported by Janaway *et al.* (2003).

A schematic diagram of an HSCCC-HPLC-PDA instrument set is shown in Fig. 4.1.

Different solvent systems such as ethyl acetate–water, ethyl acetate–methanol–water, ethyl acetate–acetic acid–water and ethyl acetate–methanol–acetic acid water were used as the two-phase solvent system for optimization of the HSCCC separation condition of baicalin and wogonoside from the Chinese medicinal plant, *Scutellaria baicalensis* Georgi (Wu *et al.*, 2005; Table 4.2). When ethyl acetate–water and ethyl acetate–methanol–water were used as the solvent systems, the target compounds mainly partitioned in the aqueous phase. Therefore, these systems were found to be unsuitable for HSCCC separation. However, when acetic acid (1%)–water was used in place of water for an ethyl acetate–methanol–water solvent system, the partition of the compounds between the upper and lower phase improved significantly. The partition coefficient (*K*) of baicalin and wogonoside and other compounds present in the crude sample in an ethyl acetate–methanol–acetic acid (1%)–water solvent system is shown in Table 4.2. The solvent systems listed in the table were also tested in HSCCC separation. The results indicated that when ethyl acetate–methanol–acetic acid (1%)–water (5:0.5:5, v/v) was used as the two-phase solvent system, baicalin and wogonoside with high purity could be separated successfully. Baicalin (58.1 mg) and wogonoside (17.0 mg) with the purity of 99.2 and 99.0%, respectively, were separated in one-step separation from 120 mg of crude sample from *S. baicalensis* Georgi. The HSCCC instrument employed in the study consisted of a three multilayer coil separation connected in a series (internal diameter (i.d.) of the tubing = 1.6 mm, total volume = 260 ml) and a 20-ml sample loop. The revolution radius was 5 cm and the β values of the multilayer coil varied from 0.5 at the internal terminal to 0.8 at the external terminal.

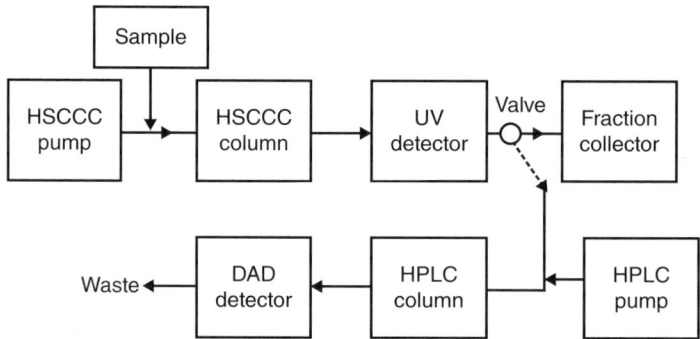

Fig. 4.1. Instrumental set-up of HSCCC–HPLC–photo diode array (PDA). (From Shi *et al.*, 2009.)

A HSCCC method for the isolation and purification of flavonoids from the Chinese medicinal plant, *Epimedium koreamum* Nakai, was developed using chloroform–methanol–water (4:3.5:2, v/v) as the two-phase solvent system (Liu *et al.*, 2005b). The HSCCC instrument consisted of a three multilayer separation column connected in a series (i.d. of the tubing = 1.6 mm, total volume = 260 ml) and a sample loop (20 ml). The revolution radius was 5 cm and the β values of the column varied from 0.5 cm at the internal terminal to 0.8 cm at the external terminal. The two-phase solvent system was selected according to the *K* values of each target component. In order to achieve efficient resolution of the target components different solvent systems were determined by HPLC (Table 4.4).

In the case of ethyl acetate–water (5:5, v/v), purity of icariside II was very low (68.2%). However, when ethyl acetate–methanol–water (5:1:5, v/v) and ethyl acetate–methanol–water (5:3:5, v/v) were used for HSCCC separation, only icariin with high purity (98%) was obtained. The purity of icariside II and epimedokoreanoside was lower than 80%. No pure compound could be isolated when ethyl acetate–methanol–water (5:3:5, v/v) was used as the two-phase solvent system. A very long time was required for separation when chloroform–methanol–water (4:2.5:2, v/v) was selected. In the case of chloroform–methanol–water (4:3:2, v/v), the three compounds were well separated but the separation time of icariside II was long and peak broadening was observed. When chloroform–methanol–water (4:3.5:2, v/v) was used, three peaks were well separated and the separation time was also acceptable. But when chloroform–methanol–water (4:4:2, v/v) was used, the three compounds could not be well separated and the purity of the compounds was also low. Therefore, chloroform–methanol–water (4:3.5:2, v/v) was used as the two-phase solvent system of HSCCC. Under optimized conditions, 11.4 mg of epimedokoreanoside I, 46.5 mg of icariin and 17.7 mg of icariside II at purity of 98.2%, 99.7% and 98.5%, respectively, as determined by HPLC, were obtained in one-step separation.

An HSCCC method was developed for the preparative separation and purification of the bioactive molecule phillyrin from *Forsythia suspensa*

Table 4.4. *K* values of the target components in different solvent systems. (From Liu *et al.*, 2005b.)

Solvent system (v/v)	*K*		
	Epimedokoreanoside I	Icariin	Icariside II
Ethyl acetate–water (5:5)	1.97	1.49	0.22
Ethyl acetate–methanol–water (5:1:5)	1.84	1.17	0.31
Ethyl acetate–methanol–water (5:2:5)	1.35	0.94	0.46
Ethyl acetate–methanol–water (5:3:5)	1.18	0.68	1.00
Chloroform–methanol–water (4:2.5:2)	2.62	4.70	8.71
Chloroform–methanol–water (4:3:2)	1.70	2.70	4.95
Chloroform–methanol–water(4:3.5:2)	1.47	2.22	3.85
Chloroform–methanol–water (4:4:2)	1.23	1.52	2.33

(Thunb.) by Li and Chen (2005). The crude phillyrin was obtained by extraction with ethanol (50%) using sonication from the dried fruits of *F. suspensa*. A two-phase solvent system for preparative HSCCC was composed of *n*-hexane–ethyl acetate–ethanol–water (1:9:1:9, v/v/v/v). Phillyrin (5.6 mg, 98.6% purity) was obtained from the crude extract (500 mg of crude extract with 1.2% phillyrin concentration) in a one-step separation.

Lu *et al.* (2007a) developed a preparative HSCCC method for the isolation and purification of the bioactive component mollugin directly from the ethanol extract of *Rubia cordifolia* using light petroleum (60–90°C) and ethanol–diethyl ether–water as the two-phase solvent system. The upper phase of light petroleum–ethanol–diethyl ether–water (5:4:3:1, v/v) was used as the stationary phase of HSCCC. Under optimum conditions, mollugin (46 mg, purity 98.5%) was obtained from crude extract (500 mg) in a single HSCCC separation.

Baicalin was separated and purified from the traditional Chinese medicinal plant, *S. baicalensis* Georgi by HSCCC (Lu *et al.*, 2003). Crude baicalin was obtained by extraction with methanol–water (70:30, v/v) from *S. baicalensis* Georgi.

The separation was performed in two steps with a two-phase solvent system composed of *n*-butanol–water (1:1, v/v). A lower phase was used as the mobile phase at a flow rate of 1.0 ml min^{-1} in the head-to-tail elution mode. A total of 37.0 mg of baicalin at 96.5% purity was yielded from 200 mg of the crude baicalin (containing 21.6% baicalin) with 86.0% recovery, as determined by HPLC analysis.

A simple and efficient HSCCC method was optimized for the preparative separation of stilbene glycosides from *Rheum tanguticum* Maxim by HSCCC (Zhao *et al.*, 2013). The solvent system developed for the separation was composed of chloroform–*n*-butanol–methanol–water (4:1:3:2, v/v/v/v). The upper phase was used as the stationary phase and the lower phase was used as the mobile phase. The flow rate was 1.8 ml min^{-1}. The apparatus was controlled at 800 rpm and 25°C and the effluent was monitored at 280 nm. Chemical constituents were analysed by HPLC and their structures were identified by ^{1}H- and ^{13}C-nuclear magnetic resonance.

4.7 Conclusion

The conventional methods of preparative separating and purifying procedures are tedious and time-consuming. An easy system capable of separating the bioactive compounds is required in drug discovery. Although HSCCC is a relatively new chromatographic technique, the versatility of HSCCC makes it an ideal choice for the isolation of bioactive natural products. The method permits the direct introduction of crude plant samples into the column without more preparation; therefore, it has been applied successfully to isolate and purify a number of natural products. Compared to preparative HPLC, HSCCC offers several advantages, such as the sample load is significantly higher and also cheap solvents are required in place of an expensive solid-phase column. Furthermore, the gentle operating conditions of HSCCC, especially the lack of active surfaces, ensure an isolation even of labile compounds (Schwarz *et al.*, 2003). As the stationary phase is liquid in nature, it is also possible to reverse the phase role even during a run using dual-mode HSCCC in order to elute all the compounds with a wide range of polarities in a short separation time. Successful separation requires a suitable two-phase solvent system having ideal partition coefficients, and its selection may account for about 90% of the entire work in HSCCC.

References

Berthod, A., Ruiz-Angel, M.J. and Carda-Broch, S. (2003) Elution-extrusion contercurrent chromatagraphy. Use of the liquid nature of the stationary phase to extend the hydrophobicity window. *Analytical Chemistry* 75, 5886–5894.

Cao, X., Huang, D., Dong, Y. and Zhao, H. (2007) Separation of aloins A and B from *Aloe vera* exudates by high speed counter current chromatography. *Liquid Chromatography and Related Technologies* 30, 1657–1668.

Chen, L.J., Games, D.E. and Jones, J. (2003) Isolation and identification of four flavonoid constituents from the seeds of *Oroxylum indicum* by high speed counter current chromatography. *Journal of Chromatography A* 988, 95–105.

Chen, L.J., Song, H., Lan, X.Q, Games, D.E. and Sutherland, I.A. (2005) Comparison of the high speed counter current chromatography instruments for the separation of the extracts of the seeds of *Oroxylum indicum*. *Journal of Chromatography A* 1063, 241–245.

Chen, Q., Yao, S., Huang, X., Luo, J., Wang, J., *et al.* (2009) Super critical fluid extraction of *Coriandrum sativum* and subsequent separation of isocoumarins by high speed counter chromatography. *Food Chemistry* 117, 504–508.

Conway, W.D. (1990) *Counter Current Chromatography: Apparatus, Theory and Application.* Wiley, New York.

Degenhardt, A. and Winterhalter, P. (2001) Isolation and purification of isoflavones from soy flour by high speed counter chromatography. *European Food Research and Technology* 213, 277–280.

Degenhardt, A., Knapp, H. and Winterhalter, P. (2000a) Separation and purification of anthocyanins by high speed counter current chromatography and screening for antioxidant activity. *Journal of Agricultural and Food Chemistry* 48, 338–343.

Degenhardt, A., Hofmann, S., Knapp, H. and Winterhalter, P. (2000b) Preparative isolation of anthocyanins by high speed counter current chromatography and application of color activity concept to red wine. *Journal of Agricultural and Food Chemistry* 48, 5812–5818.

Degenhardt, A., Engelhardt, U.H., Lakenbrink, C. and Winterhalter, P. (2000c) Preparative separation of polyphenols from tea by high speed counter chromatography. *Journal of Agricultural and Food Chemistry* 48, 3425–3430.

Degenhardt, A., Knapp, H. and Winterhalter, P. (2001) Separation of natural food colorants by high-speed countercurrent chromatography. In: Ames, J.M. and Hofmann, T. (eds) *Chemistry and Physiology of Selected Food Colorants*, ACS Symposium Series 775, American Chemical Society, Washington, DC, pp. 22–42.

Degenhardt, A., Habben, S. and Winterhalter, P. (2002) Isolation of the lignan secoisolaricire-sinol diglucoside from flaxseed (*Linum usitatissimum* L.) by high-speed countercurrent chromatography. *Journal of Chromatography A* 943, 299–302.

Destandau, E., Toribio, A., Lofosse, M., Pecher, V., Lamy, C., *et al.* (2009) Centrifugal partition chromatography directly interfaced with mass spectrometry for the fast screening and frac-tionation of major xanthones in *Garcinia mangostana*. *Journal of Chromatography A* 1216, 1390–1394.

Du, Q.Z., Li, Z.H. and Ito, Y. (2001) Preparative separation of isoflavone components in soy-beans using high speed counter current chromatography. *Journal of Chromatography A* 923, 271–274.

Hostettmann, K. (1980) Droplet counter current chromatography and its application to the pre-parative scale separation of natural products. *Planta Medica* 39, 1–18.

Hou, Z.G., Xu, D.R., Yao, S., Luo, J.G. and Kong, L.Y. (2009) An application of high speed counter current chromatography coupled with electrospray ionization mass spectrometry for separation and online detection of coumarins from *Peucedanum praeruptorum* Dunn. *Journal of Chromatography B* 2571, 2578.

Ito, Y. (1981) Efficient preparative counter current chromatography with a coil planet centrifuge. *Journal of Chromatography A* 214, 122–125.

Ito, Y. (1988) Principles and instrumentations of counter current chromatography. In: Mandava, N.B. (ed.) *Counter Current Chromatography, Theory and Practice*. Marcel Dekker, New York, p. 443.

Ito, Y. (1991) Recent advances in counter current chromatography. *Journal of Chromatography A* 538, 3–25.

Ito, Y. (2005) Golden rules and pitfalls in selecting optimum conditions for high speed counter current chromatography. *Journal of Chromatography A* 1065, 145–168.

Janaway, L., Hawes, D., Ignatova, S., Wood, P. and Sutherland, I.A. (2003) A new small coil volume CCC instrument for direct interfacing with MS. *Journal of Liquid Chromatography and Related Technologies* 26, 1345–1354.

Jia, J., Chen, T., Wang, P., Chen, G., You, J., *et al.* (2012) Preparative separation of methylswertianin, swerchirin and decussatin from Tibetan medicinal plant *Swertia mus-sotii* using high speed counter current chromatography. *Phytochemical Analysis* 23, 332–336.

Kim, Y.S. and Ha, Y.W. (2009) Preparative isolation of six major saponins from *Platycodi radix* by high speed counter current chromatography. *Phytochemical Analysis* 20, 207–213.

Leister, W., Strauss, K., Wisnoski, D. and Zhao, Z. (2003) Development of a custom high throughput preparative liquid chromatography/mass spectrometer platform for the prepara-tive purification and analytical analysis of compound libraries. *Journal of Combinatorial Chemistry* 5, 322–329.

Li, A., Sun, A. and Liu, R. (2005) Preparative isolation and purification of costunolide and dehy-drocostuslactone from *Aucklandia lappa* Decne by high speed counter chromatography. *Journal of Chromatography A* 1076, 193–197.

Li, H.B. and Chen, F. (2005) Preparative isolation and purification of phillyrin from the medi-cinal plant *Forsythia suspensa* by high-speed counter-current chromatography. *Journal of Chromatography A* 1083, 102–105.

Li, H.B. and Chen, F. (2009) Separation and purification of epicedia A, B, C and icariin from medicinal herb *Epimedium brevicornum* Maxim by dual-mode HSCCC. *Journal of Chromatographic Science* 47, 337–340.

Liu, R.M., Chu, X., Sun, A.L. and Kong, L.Y. (2005a) Preparative isolation and purification of flavonoids from the Chinese medicinal herb *Evodia rutaecarpa* (Juss.) Benth by high speed counter current chromatography. *Journal of Chromatography A* 1074, 139–144.

Liu, R., Li, A., Sun, A., Cui, J. and Kong, L. (2005b) Preparative isolation and purification of three flavonoids from the Chinese medicinal plant *Epimedium koreanum* Nakai by high-speed counter-current chromatography. *Journal of Chromatography A* 1064, 53–57.

Liu, R.M., Wu, S.A. and Sun, A.L. (2008) Separation and purification of four chromones from *Radix saposhnikoviae* by high speed counter current chromatography. *Phytochemical Analysis* 19, 206–211.

Lu, H.T., Jiang, Y. and Chen, F. (2003) Application of high-speed counter-current chromatography to the preparative separation and purification of baicalin from the Chinese medicinal plant *Scutellaria baicalensis*. *Journal of Chromatography A* 1017, 11–123.

Lu, J.J., Wei, Y. and Yuan, Q.P. (2007b) Preparative separation of gallic acid from Chinese traditional medicine by high speed counter current chromatography and followed by liquid chromatography. *Separation and Purification Technology* 55, 40–43.

Lu, Y.B., Sun, C.R. and Pan, Y.J. (2006) A comparative study of upright counter current chromatography and high performance liquid chromatography for preparative isolation and purification of phenolic compounds from *Magnoliae officinalis*. *Journal of Separation Science* 29, 351–357.

Lu, Y., Liu, R., Sun, C. and Pan, Y. (2007a) An effective high-speed countercurrent chromatographic method for preparative isolation and purification of mollugin directly from the ethanol extract of the Chinese medicinal plant *Rubia cordifolia*. *Journal of Separation Science* 30, 1313–1317.

Oka, H., Harada, K., Suzuki, M. and Ito, Y. (2000) Separation of spiramycin components using high-speed counter-current chromatography. *Journal of Chromatography A* 903, 93–98.

Oka, H., Harada, K., Ito, Y. and Ito, H. (2001) High-speed counter-current chromatography separation and purification of resveratrol and piceid from *Polygonum cuspidatum*. *Journal of Chromatography A* 907, 343–346.

Peng, A.H., Li, R., Hu, J., Chen, L.J., Zhao, X., et al. (2008) Flow rate gradient high speed counter chromatography separation of five diterpenoids from *Triperygium wilfordii* and scale up. *Journal of Chromatography A* 1200, 129–135.

Peng, J., Fan, G., Hong, Z., Chai, Y. and Wu, Y. (2005) Preparative separation of isovitexin and isoorientin from *Patrinia vilosa* Juss by high speed counter current chromatography. *Journal of Chromatography A* 1074, 111–115.

Schwarz, M., Hillebrand, S., Habben, S., Degenhardt, A. and Winterhalter, P. (2003) Application of high-speed counter current chromatography to the large scale isolation of anthocyanins. *Biochemical Engineering Journal* 14, 179–189.

Sethi, P.D. (ed.) (2013) *HPTLC-Quantitative Analysis of Pharmaceutical Formulations*. CBS Publishers and Distributors, New Delhi.

Shi, S., Jiang, D., Zhao, M. and Tu, P. (2007) Preparative isolation and purification of triterpene saponins from *Clematis mandshurica* by high-speed counter-current chromatography coupled with evaporative light scattering detection. *Journal of Chromatography B* 852, 679–683.

Shi, S., Zhang, Y., Zhao, Y. and Huang, K. (2008) Preparative isolation and purification of three flavonoid glycosides from *Taraxacum mongolicum* by high-speed counter-current chromatography. *Journal of Separation Science* 31, 683–688.

Shi, S., Zhou, H., Zhang, Y., Zhao, Y., Huang, K., et al. (2009) A high speed counter chromatography HPLC-DAD method for preparative isolation and purification of two polymethoxylated flavones from *Taraxum mongolicum*. *Journal of Chromatographic Science* 47, 349–353.

Srivastava, M. (ed.) (2011) *High Performance Thin Layer Chromatography (HPTLC)*. Springer, Heidelberg, New York.

Stahl, E. (1969) *Thin Layer Chromatography*. Springer, Berlin.

Sun, Y.S., Zhu, H.F., Wang, J.H., Liu, Z.B. and Bi, J.J. (2009) Isolation and purification of salvi-anolic acid A and savianolic acid B from *Salvia miltiorrhiza* by high-speed counter current chromatography and comparison of their antioxidant activity. *Journal of Chromatography B* 877, 733–737.

Tang, Q.F., Liu, J.H., Xue, J., Ye, W.C., Zhang, Z.J., *et al.* (2008) Preparative isolation and puri-fication of two isomeric diterpenoid alkaloids from *Aconitum coreanum* by high speed counter current chromatography. *Journal of Chromatography B* 872, 181–185.

Tang, Q., Wang, Y., Chen, M., Zhang, Q., Fan, C., *et al.* (2013) Application of high-speed counter chromatography preparative separation of flavones C-glycosides from *Lophatherum gracile*. *Separation Science and Technology* 48, 1906–1912.

Urbain, A., Marston, A., Batsuren, D., Purev, O. and Hostettmann, K. (2008) Preparative iso-lation of closely related xanthones from *Gentianella amarelle* ssp. *acuta* by high-speed countercurrent chromatography. *Phytochemical Analysis* 19, 514–519.

Wang, Y.F. and Liu, B. (2007) Preparative isolation and purification of dicaffeoylquinic acids from the *Ainsliaea fragrans* Champ by high-speed countercurrent chromatography. *Phytochemical Analysis* 18, 436–440.

Wei, Y., Zhang, T.Y. and Ito, Y. (2004) Preparative isolation of osthol and xanthotoxol from common cnidium fruit (Chinese traditional herb) using stepwise elution by high-speed countercurrent chromatography. *Journal of Chromatography A* 1033, 373–377.

Winterhalter, P., Sefton, P.J. and Williams, P.J. (1990) Two-dimensional GC-DCCC analysis of glycoconjugates of monoterpenes, norisoprenoids and shikimate derived metabolites from Riesling wine. *Journal of Agricultural and Food Chemistry* 38, 1041–1048.

Wu, S., Sun, A. and Liu, R. (2005) Separation and purification of baicalin and wogonoside from the Chinese medicinal plant *Scutellaria baicalensis* Georgi by high-speed counter-current chromatography. *Journal of Chromatography A* 1066, 243–247.

Xin, X.L., Yang, Y., Zhong, J., Aisa, H.A. and Wang, H.Q. (2009) Preparative isolation purifi-cation of sobenzofuranone derivatives and saponins from seeds of *Nigella glandulifera* Freyn by high-speed countercurrent chromatography combined with gel filtration. *Journal of Chromatography A* 1216, 4258–4262.

Yang, C.H., Tang, Q.F., Liu, J.H., Zhang, Z.H. and Liu, W.Y. (2008) Preparative isolation and purification of phenolic acids from *Smilax china* by high-speed counter-current chromatog-raphy. *Separation and Purification Technologies* 61, 474–478.

Yao, S., Liu, R.M., Huang, X.F. and Kong, L.Y. (2007) Preparative isolation and purification of chemical constituents from the root of *Adenophora tetraphlla* by high-speed counter-current chromatography with evaporative light scattering detection. *Journal of Chromatography A* 1139, 254–262.

Yuan, E.D., Liu, B.G., Ning, Z.X. and Chen, C.G. (2009) Preparative separation of flavonoids in *Adinandra nitida* leaves by high-speed counter-current chromatography and their effects on human epidermal carcinoma cancer cell lines. *Food Chemistry* 115, 1158–1163.

Zhao, W.H., Cao, C.C., Ma, X.F., Bai, X.Y. and Zhang, Y.X. (2007) Isolation of 1, 2, 3, 4, 6-penta-o-galloyl-beta-D-glucose from *Acer truncatum* Bunge by high-speed counter-current chro-matography. *Journal of Chromatography B* 850, 523–527.

Zhao, X.H., Han, F., Li, Y.L. and Yue, H.L. (2013) Preparative isolation and purification of three stilbene glycosides from the Tibetan medicinal plant *Rheum tanguticum* Maxim. Ex Balf. by high-speed counter-current chromatography. *Phytochemical Analysis* 24, 171–175.

Zhou, T.T., Chen, B., Fan, G.R., Chai, Y.F. and Wu, Y.T. (2006) Application of high-speed counter-current chromatography coupled with high-performance liquid chromatography diode array detector for the preparative isolation and purification of hyperoside from *Hypericum perforatum* with online purity monitoring. *Journal of Chromatography A* 1116, 97–101.

Zhou, T.T., Zhu, Z.Y., Wang, C., Fan, G.R., Peng, J.Y., *et al.* (2007) Online purity monitoring in high-speed counter-current chromatography: application of HSCCC-HPLC-DAD for the preparation of 5-HMF, neomangiferin from *Anemarrhena asphodeloides* Bunge. *Journal of Pharmaceutical and Biomedical Analysis* 44, 96–100.

Zhu, L.C., Yu, S.J., Zeng, X.N., Fu, X. and Zhao, M. (2008) Preparative separation and purification of five anthraquinones from *Cassia tora* by high-speed counter-current chromatography. *Separation and Purification Technology* 63, 665–669.

5

Chromatography Techniques II: High-performance Liquid Chromatography and Ultra-performance Liquid Chromatography

5.1 Introduction

Chromatography is an analytical technique based on the separation of compounds due to differences in their structure or composition. In general, chromatography involves moving a sample through the system over a stationary phase. Different affinities and interactions of molecules in the sample with the stationary support lead to separation of molecules. Sample components having stronger interactions with the stationary phase move more slowly through the column in comparison to components having weaker interactions. On this basis, different compounds can be separated from each other as they move through the column. High-performance liquid chromatography (HPLC) is a type of chromatography used to separate and quantify the compounds that have been dissolved in solution. HPLC is presently the most widely used method of qualitative and quantitative analysis in drug discovery from plant sources. Since the early 1980s, HPLC has been recognized as the most versatile technique for separating natural products directly in crude extract without the need for complex sample preparation.

Typically, a HPLC system consists of the following components: (i) solvent reservoir; (ii) pump system; (iii) sample injection system; (iv) detector; and (v) a computer serving as the data station for the detector information, as well as a way to control and automate the HPLC pump (Fig. 5.1).

5.1.1 Pump

More sophisticated instruments consist of two to four pumps, as well as more than one detector for qualitative and quantitative characterization of the sample. Pump systems are used to maintain the continuous flow of the mobile phase. The pressure generally reaches 1500–2000 bars.

  © Satyanshu Kumar 2016. *Analytical Techniques for Natural Product Research* (S. Kumar)

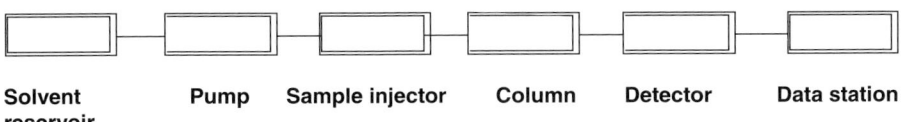

Solvent   Pump   Sample injector   Column   Detector   Data station
reservoir

Fig. 5.1. Schematic diagram of a high-performance liquid chromatography (HPLC) system.

As in some cases, very high pressure is unavoidable; modern instruments can withstand up to 6000 atmospheres. It is possible to change solvent composition throughout the chromatography using gradient programming.

5.1.2 Column

The column used is the heart of the HPLC system as separation takes place in the column. The development of columns with different phase chemistry, especially reverse-phase columns, has enabled the separation of almost any type of natural product. Columns differing in stationary phase pore size are initially screened for their potential to resolve the components, with special attention to separation efficiency, peak shape and separation time.

The introduction of very high pH-stable phases, sub-2-µm particles and monolith columns has improved the performance of HPLC systems considerably in terms of speed, resolution and reproducibility (Nguyen *et al.*, 2006; Wolfender, 2009).

Columns used in HPLC analysis are made of a stationary phase with small particle size. Because of this, a high back pressure is generated when the solvent or mobile phase passes through the column. HPLC columns differ in terms of stationary phase (C-8, C-12, C-18), particle size (3–5 µm), pore size (80–200 Å) and length (10–25 cm). The most popular packing materials for a HPLC column are silica based. Octadecyl silica (ODS-silica) is the most commonly used material, and contains a C_{18} coating. Other materials containing C_1, C_2, C_4, C_6, C_8 and C_{22} coatings are also used. For the purification of specific compounds, different types of chemical moieties have also been designed. Monolithic, zirconia and polymer packing material-based columns are also available. Silica-based particles have good mechanical strength, but can suffer from a number of disadvantages including a limited pH range and tailing of basic analytes. Polymeric columns can overcome pH limitations. Hybrid columns incorporating carbons in the form of methyl groups are mechanically strong with high efficiency and operate over an extended pH range. A second-generation hybrid column such as bridged ethane hybrid (BEH) was developed by bridging the methyl group in the silica matrix (Swartz, 2005).

For method development, different HPLC columns are screened with the intention of obtaining better separation efficiency, peak shape and retention time. By using a column of smaller particle size, the flow rate could be reduced. Further, increase in column temperature also leads to considerable reduction in the separation time. Elevated temperature provides the dual advantages of lowering viscosity and increasing mass transfer by increasing the diffusivity of the analytes (Neue and Mazzeo, 2001).

5.1.3 Solvent or mobile phase

Polarity and composition of the mobile phase affect the separation of components. Solvents of varying polarity are used for HPLC method development. In normal-phase HPLC, usually the solvent is non-polar and in reversed-phase HPLC (RP-HPLC), the solvent is normally a mixture of water and a polar organic solvent. Separation is performed mostly in reverse-phase chromatography with acetonitrile–methanol–water or a combination of more than two solvents in isocratic or gradient elution mode. Mobile-phase pH affects the ionization of the functional group. This in turn affects ion pair formation and retention of the solutes. Modifiers having low ultraviolet (UV) cut-offs, such as phosphate buffer, trifluoroacetic acid, acetic acid, etc., are also used along with organic solvent.

5.1.4 Sample introduction

The chromatographic process begins with the injection of the sample at the injector end of the column before the chromatography system is equilibrated with the mobile phase. Separation of components takes place as the analytes and mobile phase are pumped through the column. As the mobile phase can be adjusted to the changing polarity of the mixture in gradient elution mode, a much larger number of compounds can be separated in a mixture. Additionally, in gradient elution mode, the chromatographic peaks become sharper in a much shorter time. An isocratic flow system is also chosen to minimize baseline drift. Also, isocratic elution is more simple, precise, accurate and stable than a gradient system, and thus is more suitable for quality control and routine analysis (Sun and Su, 2000). Components eluting from the column appear as peaks on the data station.

5.1.5 Detector

The choice of an appropriate detector is crucial because of the diversity in the physico-chemical properties of natural products. The refractive index detector is the most simple and least expensive detector. This detector has been useful for detecting compounds such as carbohydrates and polymers, but it lacks sensitivity and is very susceptible to changes in ambient temperature, pressure and flow rate. Furthermore, it cannot be used for gradient elution. Because of these limitations, the refractive index detector has been replaced by other detectors.

The UV detector is the most simple and most widely used among all HPLC detectors. The relationship between the intensity of light transmitted through the detector cell and solute concentration is given by Beer's law. The magnitude of the extinction coefficient of an analyte at a given wavelength controls the sensitivity of the detection. This detector cannot be used for natural products lacking chromophores. Also, mobile-phase constituents having high UV cut-off wavelenths should be avoided because they might prevent the detection of analytes with weak chromophores (Ozkan, 2007). The photodiode array detector is a multiple wavelength UV detector and is also more versatile than the fixed wavelength detector. The UV spectra of the constituent can be monitored even during the separation. For those compounds possessing a weak chromophore, an unspecific wavelength close to 200 nm is used for detection (Ganzera *et al.*, 2004).

Fluorescence detectors afford greater sensitivity. Here, the molecular absorption of photons triggers the emission of another photon with a longer wavelength. This difference in wavelengths, that is, absorption versus emission, provides more selectivity. Fluorescent light is measured against a very low light background. Compared with UV, fluorescence detectors have scarcely been used for the detection of natural products. Most of the applications are related to very sensitive detection of aflatoxins in food because this class of natural product contains natural fluorescence (Jaimez *et al.*, 2000). The detection of natural products that do not fluoresce naturally has been achieved successfully after the addition of fluorescent tags (Kristl *et al.*, 2005). Similar to fluorescence, chemiluminescence can be defined as the emission of light from a molecule or atom in an electronically excited state produced by a chemical reaction at an ordinary temperature without any associated generation of heat. Chemiluminescent nitrogen detection is a relatively new technology for the detection of nitrogen-containing molecules, including a broad range of pharmaceuticals with very high sensitivity up to the femtogram range (Li *et al.*, 2003). As many compounds are not chemiluminescent, they can only be detected by chemiluminescence after being derivatized (Ohba *et al.*, 2002). The analysis of flavonoids (limit of detection 3 ng ml^{-1}) in phytopharmaceuticals containing *Hippophae rhamnoides* has been reported. The procedure was based on the chemiluminescent enhancement of the flavonols by a cerium (IV) rhodamine 6G system in a sulfuric acid medium (Zhang and Cui, 2005). Natural products with electrochemical activity are common. Natural products with electroactive groups are readily measurable and detectable by liquid chromatography with electrochemical detection. This technique can be applied to a large number of analytes in either the oxidation or reduction mode. Functional groups such as phenols, aromatics, amines, thiols, quinolones, etc., are sensitive to oxidation, while functionals groups such as olefins, esters, ketones, aldehydes, ethers, quinones, etc., are compatible with reduction. Cells of an electrochemical detector consist of three electrodes, the working, counter and reference electrodes. Although two designs exist for these electrodes, they can be aligned in several different geometries. The eluent is directed through the electrode in coulometric systems, while the eluent passes over the electrode in amperometric systems. Electrochemical detection is usually performed with maintaining the potential of the working electrode at a fixed value relative to the potential of the working electrode at a fixed value, which is measured by the reference electrode. The electrochemical reaction is driven by the fixed potential difference applied between the working and the reference electrodes. The current produced is measured as a function of the elution time, allowing detection at the picomolar level. The electrochemical detector differs from other detectors because it alters the sample; however, it is less selective than the fluorescent detector. This method of detection is inexpensive, sensitive and widely accepted. The accessible potential range, the number of compounds that are active in this range and half-width of the individual signals determine the selectivity of the electrochemical detector (Ozkan, 2007). Multi-step potential time wave forms involving pulse techniques are also available for electrochemical detection. Here, uniform and reproducible electrode activity is maintained, realizing amperometric or coulometric detection (LaCourse and Modi, 2005). Proper use of electrochemical detectors requires knowledge of redox reactions and their dependence on mobile-phase composition. Study of the electrochemical reactions mechanism of natural products may supplement the establishment of their interaction with living cells (Guo *et al.*, 1997).

The evaporative light scattering detector (ELSD) was introduced in 1966 (Ford and Kennard, 1966). Any analyte that is less volatile than the mobile phase, regardless of the optical, electrochemical or other analyte properties, can be detected using ELSD. The eluent is nebulized using a flow of nitrogen, and the resulting aerosol is transported through a heated drift tube. Volatile components and solvents are evaporated in the drift tube. The solid fraction remaining is subsequently introduced into a detection cell. A light beam is directed on to the particles, which causes scattering of the incident light detected by a photodiode or photomultiplier. Nebulizer gas flow and drift tube temperatures are the most important parameters affecting the ELSD signal response. Gas flow rate influences the droplet size of the column effluent before evaporation occurs. Higher flow rate results in the formation of smaller aerosol droplets and less scattering of light, causing lower sensitivity but allowing a stable baseline. Drift tube temperature facilitates the evaporation of the nebulized aerosol so that the light scattering response of the non-volatile solute can be determined exclusively. In comparison to the semi-volatile mobile phase, a higher temperature is needed for non-volatile analytes, as well as a mobile phase with high aqueous content. The new generation ELSD detector is able to vaporize eluent at a low temperature, thereby facilitating the detection of semi-volatile analytes (Megoulas and Koupparis, 2005; Guillarme *et al.*, 2008). In contrast to the UV detector, which is concentration dependent, the ELSD is mass dependent. Therfore, theoretically, ELSD generates a similar response for equal amounts of mass present, and thus a universal response factor. The ELSD response also depends on the volatility of the compound and the mobile-phase composition in the case of gradient elution, which makes such quantification procedures experimentally inaccessible. Thus, the interplay of several factors leads to a non-linear response. This disadvantage renders linear regression for calibration curves inaccurate (Vervoort *et al.*, 2008). ELSD has been widely used as an alternative to the UV detector for analytes lacking chromophores or with weak chromophores (Wolfender, 2009). ELSD techniques are frequently used in combination with UV detection.

Dixon and Peterson (2002) introduced a charged aerosol detector as a new technology for the universal HPLC detector. The first step of the charged aerosol detector is similar to ELSD, and the dried particle stream is charged with a corona discharge needle. The current generated is measured by an electrometer (Vervoort *et al.*, 2008).

The flame ionization detector (FID), which is a general detector of gas chromatography, has been applied to HPLC (Guillarme *et al.*, 2008). HPLC-FID coupling implies working with 100% water as the mobile phase; however, water is a very weak mobile phase in reverse-phase liquid chromatography at room temperature. For selected compounds with appropriate stationary phases, separations have been achieved with pure water at ambient temperature (Foster and Synovec, 1996).

5.1.6   Identification of peaks

The identification and quantification of peaks is done when the same retention time and peak areas of the repetitive injection of the standard samples are obtained. Peaks are identified by comparing their retention time and UV absorption bands with the retention time and UV absorption bands of standard compounds. Further, the UV absorption bands obtained by the HPLC detectors are compared with the literature values of absorption bands.

The calibration curve is usually obtained by injecting known amounts of samples and measuring the peak height or peak area. The concentration of a compound is determined by plotting the peak area or height against the concentration of the standard component. Both peak area and peak height are proportional to the concentration of well-resolved peaks.

Three different calibration methods can be applied for the quantification of components. The external standard calibration method is the simplest of the three methods. The accuracy of this method is dependent on the reproducibility of the injection volume. Here, different known concentrations of standard solution are prepared and a fixed volume of each concentration is injected. The peak area or height is plotted against the concentration to obtain a linear equation. From the linear regression equation, the concentration of the component in an unknown sample can be obtained. The internal standard method is used to obtain the most accurate and precise quantification. In this method, an equal amount of an internal standard, a compound that is not present in the sample, is added to both the sample and standard solutions. Quantification is done by using the ratios of peak height or peak area of the compound to the internal standard. The internal standard should be stable, have a similar retention time and does not interfere with the any of the sample components. The standard addition method is also used for the construction of the calibration curve.

5.2   Basic Concepts of HPLC Method Development

Some basic concepts for utilization of the practical applications of HPLC are described below.

5.2.1   Retention time (t_r)

Retention time is defined as the time elapsed between the injection of the sample and the appearance of the maximum peak response of the eluted sample. It is used as a parameter for identification. Chromatographic retention times are characteristics of the compounds, but are not unique. Coincidence of the retention times of analytes and a reference analyte can be used as one of the criteria for identification. Further, consistency of retention time confirms the repeatability of the method.

5.2.2   Capacity factor (k')

The retention time of an analyte relative to column dead volume is measured in terms of capacity factor. Dead volume or void volume of a HPLC column is defined as the volume at which an unretained compound elutes. It is defined as:

$$k' = (V_r - V_o)/V_o \text{ or } (t_r - t_o)/t_o, \tag{5.1}$$

where V_r = retention volume of analyte; V_o = dead volume or void volume; t_o = column dead volume; and t_r = retention time of analyte.

Variation in mobile-phase composition, change in the surface chemistry of the column due to ageing of the column and temperature variation usually affect the capacity factor. The value of k' in the range of 2–8 corresponds to the analyte band. Inadequate resolution gives a lower value, whereas peak broadening leads to a higher value of the capacity factor.

5.2.3   Selectivity (α)

Selectivity and sensitivity depend on the spectral properties of the solute and the performance of the detector (Engelhardt and Siffrin, 1997). It is the relative retention of two peaks in a chromatogram and is defined as:

$$\alpha = (k'_2/k'_1), \tag{5.2}$$

where k_2' = the capacity factor of analyte 2 and k_1' = the capacity factor of analyte 1. Generally, an α value between 1.05 and 2.00 is considered adequate.

5.2.4 Number of theoretical plate (N_s)

The number of the theoretical plate (N_s), also called column efficiency, relates chromatographic separation to the theory of distillation and measures column efficiency.

It is calculated using the following equation:

$$N_s = 16\,(t_r/w)^2 = 5.54\,(t_r/w_{1/2})^2, \tag{5.3}$$

where t_r = the retention time of the analyte; w = bandwidth; $w_{1/2}$ = bandwidth at half height.

The N_s value can be used as an indicator of the column efficiency, as well as the total efficiency of the liquid chromatography system in routine analysis. A significant decrease in the N_s value over time may indicate a decrease in column efficiency. The N_s value depends on the analyte characteristics as well as on the operating conditions, such as flow rate and temperature of the mobile phase, quality of packing, the uniformity of packing within the column and the thickness of the stationary phase.

5.2.5 Resolution (R_s)

Resolution, R_s, is the separation of two analytes in a mixture and indicates the quality of separation. Resolution is measured using the following equation:

$$R_s = 2\,[(t_{r2} - t_{r1})/(w_1 + w_2)] = 1.18\,[(t_{r2} - t_{r1})/(w_{1(1/2)} + w_{2(1/2)})], \tag{5.4}$$

where t_{r2} = retention time of analyte 2; t_{r1} = retention time of analyte 1; w_1 = bandwidth of peak 1; w_2 = bandwidth of peak 2; $w_{1(1/2)}$ = bandwidth at half height for peak 1; $w_{2(1/2)}$ = bandwidth at half height for peak 2.

Resolution is a function of a number of theoretical plates (N_s), selectivity (α) and capacity factor (k') and is also expressed as:

$$R_s = (\sqrt{N_s})/4 \times (\alpha - 1)/\alpha \times (k'/k' + 1). \tag{5.5}$$

For a given stationary phase and mobile phase, N needs to be specified so that: (i) it is ensured that closely eluting compounds are resolved from each other; (ii) to establish the general resolving power of the system as well as the method developed; and (iii) to ensure that internal standards are resolved from the analyte. An R_s value not less than 1.5 is strongly recommended for quantitative analysis. Quantification with R_s less than 1 is generally considered as less precise.

5.2.6   Symmetry or tailing factor (A_s)

This is a measure of peak symmetry. The value of A_s is unity for the perfectly symmetrical peaks and it increases further as tailing increases. In a case where the A_s value is less than unity, precision becomes less reliable. An A_s value between 0.9 and 1.3 is preferable, to ensure accuracy and precision of the method developed. It increases under inadequate HPLC conditions and with decreasing column efficiency.

5.2.7   Sensitivity and signal-to-noise ratio (*S/N*)

Sensitivity is the lowest detectable level of a component in a chromatographic separation and is dependent on the signal-to-noise ratio (*S/N*) for a given detector. The value of the *S/N* ratio is a useful system suitability parameter. Since the maximum sensitivity is obtained only at the maximum absorption of the solute, variable wavelength detectors or diode array detectors are preferred (Engelhardt and Siffrin, 1997).

The Van Deemter equation is an empirical formula that describes the relationship between band broadening (plate height) and flow rate (linear velocity) (Willard *et al.*, 1986).

5.2.8   System suitability

System suitability tests are an integral part of the validation of the chromatographic method. These tests are used to verify that the chromatographic system is adequate for the intended analysis. These tests are based on the concept that the equipment electronics, analytical operations and samples analysed constitute an integral system that can be evaluated as such. The following factors may affect chromatographic behaviour:

1. Composition, ionic strength, temperature and pH of the mobile phase.
2. Flow rate, column dimension, column temperature and pressure.
3. Stationary-phase characteristics including type of chromatographic support (particle based or monolithic), particle or macrospore size, porosity and specific surface area.
4. Reverse phase and other surface modification of the stationary phases, the extent of chemical modification (as expressed by end capping, carbon loading, etc.).

System suitability tests are valuable and have been accepted because reliable and reproducible chromatographic results are based on a wide range of specific parameters (Synder *et al.*, 1997).

5.2.9   Peak purity

Peak purity of all peaks of interest is investigated by studying the photo diode array (PDA) data of the corresponding peaks (Fig. 5.2). Further, chromatograms

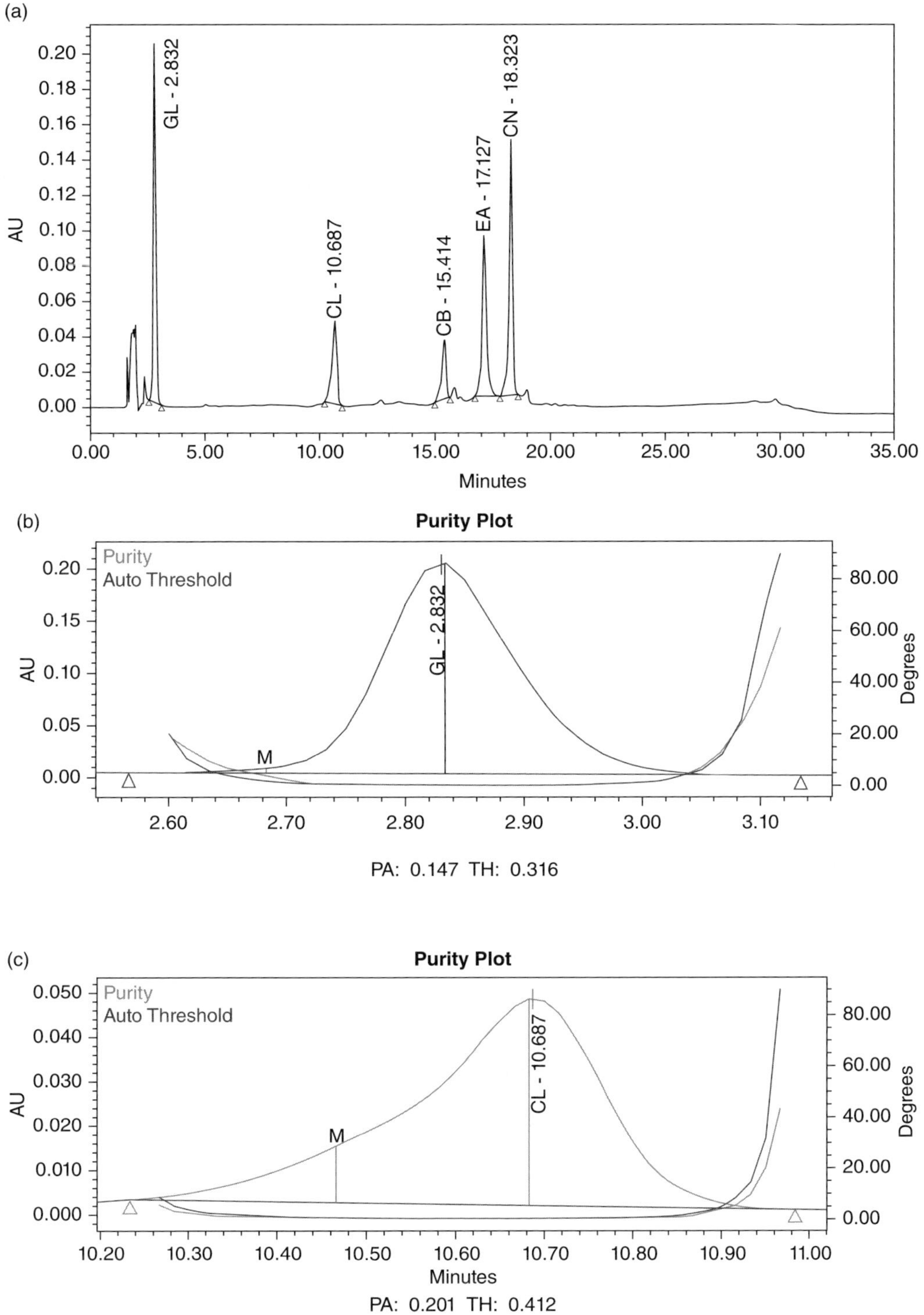

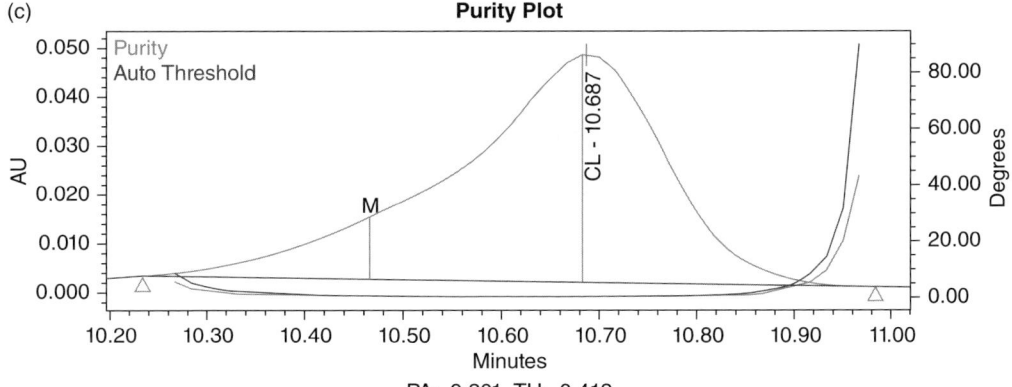

Fig. 5.2. Peak purity of standard mixture gallic acid (GL), corilagin (CL), chebulagic acid (CB), ellagic acid (EA) and chebulinic acid (CN). PA = purity angle, TH = peak threshold.

(d)

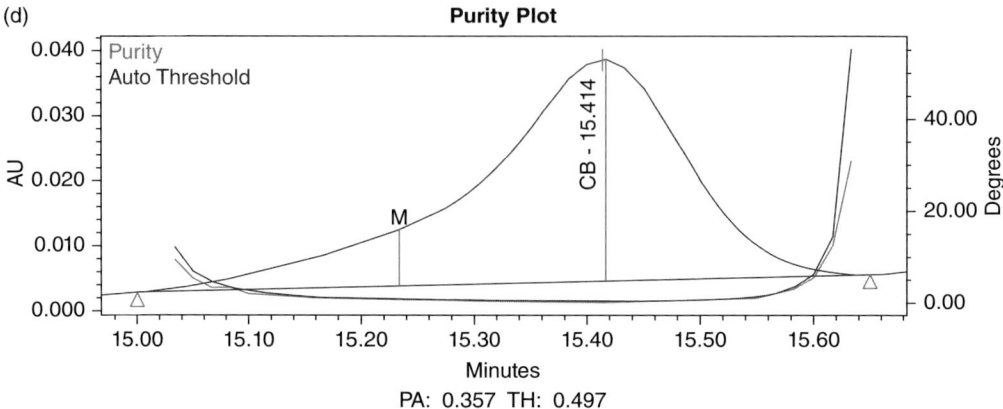

PA: 0.357 TH: 0.497

(e)

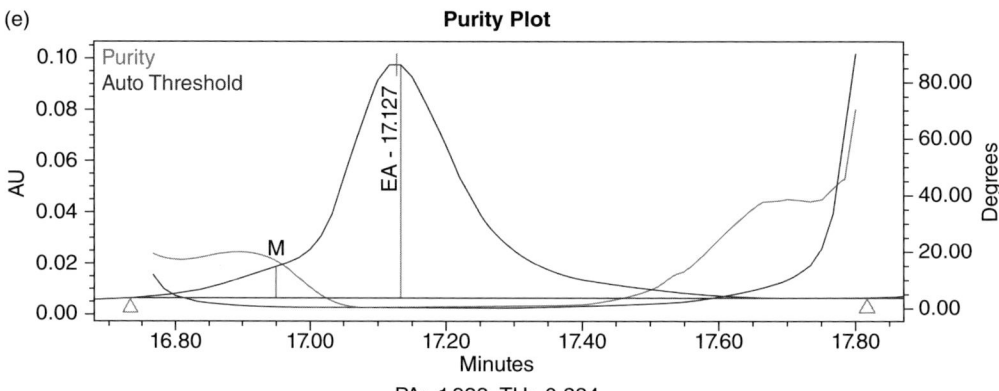

PA: 1.923 TH: 0.334

(f)

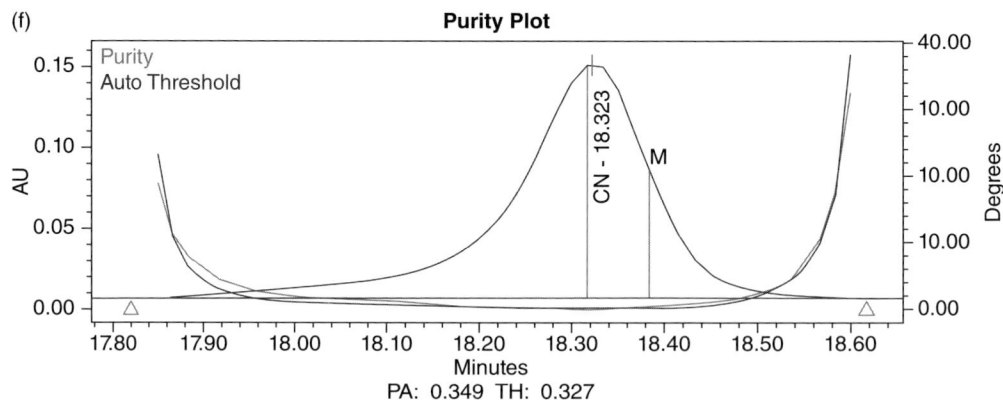

PA: 0.349 TH: 0.327

Fig. 5.2. Continued.

are checked for the appearance of any extra peaks. Peak purity is measured in terms of peak purity angle and purity threshold values. Dey *et al.* (2012) reported that for spectrally homogeneous and pure peaks of glycyrrhizic acid, a polyherbal preparation containing aqueous root extract of

Glycyrrhiza glabra, purity angle and purity threshold values were 0.357 and 0.438, respectively.

5.3 Method Validation

RP-HPLC has been the technique of choice for both pharmaceutical and bioanalytical liquid chromatography–mass spectrometry–mass spectrometry (LC-MS-MS) analysis because of the high efficiency of the separations, the compatibility of the mobile phase with biological and lipophilic samples and the easy interfacing with a variety of detectors (Wilson *et al.*, 2005; Rainville *et al.*, 2007).

Separation of the compounds present in the extract is achieved by meticulous assessment, as well as the selection of all separation-relevant parameters such as stationary phase and elution conditions. Preliminary experiments are carried out to optimize the experimental parameters affecting both the chromatographic separation of the target compounds in the column selected and their detection by UV or PDA.

After optimization of the HPLC parameters, the developed method is validated with regard to its specificity, linearity, accuracy and precision using the International Conference on Harmonisation (ICH) guidelines. The validity of an analytical method can be verified by establishing several analytical and statistical parameters. As the major components could be present in several closely related species, therefore, a validated analytical procedure for the separation and quantification of all major activity related compounds is required. Such a method would not only facilitate the standardization of commercial products but also increase the significance of future pharmacological studies, since the chemically well-defined samples can be utilized (Ganzera *et al.*, 2004). The International Conference on Harmonization of Technical Requirements for Registration of Pharmaceuticals for Human Use (ICH, 2005) guidelines provide details of method validation.

The developed HPLC method is validated in the terms of the following parameters: (i) specificity; (ii) selectivity; (iii) linearity; (iv) limit of detection and limit of quantification; and (v) precision expressed as repeatability of retention time, peak area and recovery.

5.3.1 Specificity

The ability of the analytical method to assess the analyte unambiguously in the presence of other components (impurities and degradants) can be demonstrated by evaluating specificity (Dongre *et al.*, 2007).

Specificity requires the method to provide separation of the analyte from process impurities, degradation, excipients and other analytical artefacts. The specificity of the method can be delineated by the resolution of the integrity of the peak determined by multiple UV wavelength detection;

that is, photodiode array detector or ratio between two wavelengths. However, it may not always be reliable, as many degradants and metabolites may have chromophores similar to those of the analyte. The specificity of the method developed is confirmed by the reliability of the compound peaks corresponding to the standard compounds in the sample.

Further, the robustness of the method is verified by carrying out experiments on instruments of different make. The reliability of the method is determined by making small changes in chromatographic conditions, such as the composition of the mobile phase (+5%), pH (+0.1%), etc. (Dey *et al.*, 2012).

5.3.2 Linearity

The linearity of the method is checked for the standard compounds with their respective calibration curves in the different concentration range, depending on the limit of detection and the limit of quantification. The regression equation and correlation coefficients are obtained with a minimum of six replicate analyses corresponding to each concentration of a minimum of six concentration levels. The value of the correlation coefficients should be greater than 0.99.

5.3.3 Limit of detection (LOD) and limit of quantification (LOQ)

Limit of detection (LOD) is defined as the lowest amount of sample concentration that can be detected (signal-to-noise ratio = 3), and the limit of quantification (LOQ) is defined as the lowest amount of sample concentration that can be determined quantitatively with suitable precision and accuracy (signal-to-noise ratio = 10). Both LOD and LOQ are also calculated according to the following formula:

$$\text{LOD} = 3.3 \times (\sigma/s), \text{LOQ} = 10 \times (\sigma/s), \tag{5.6}$$

where σ = standard deviation of the blank response and s = the slope of the calibration curve.

5.3.4 Precision and accuracy

The precision of the method is validated by both intraday and interday variations. To determine the intraday variation, six assays are carried out on the same sample at different times during the day. Interday variation is determined by analysing on the next day. Intraday and interday precision of the assay are expressed as relative standard deviation (RSD). RSD is calculated as RSD = standard deviation (σ)/mean and is taken as a measure of precision.

Accuracy of the developed HPLC method is determined by spiking the sample with a known amount of the standards at three concentration levels (low, medium and high) in the calibration range. Spiked samples are analysed under optimized conditions and recovery is obtained.

5.3.5 Lack-of-fit test

The lack-of-fit test is the proper test for linearity. There is always a risk that a regression model is a poor approximation of the true functional relationship between a set of data X and Y. This test is performed to support the adequacy of the regression model and is based on the hypothesis that the linear model fits the data (H_0) adequately or the linear model does not fit the data (H_1). The test involves the partioning of the residual sum of squares (SE) into the sum of squares of the lack of fit to the regression model. If the null hypothesis is rejected, the model must be abandoned since no linear relationship exists between the two variables. If the null hypothesis cannot be rejected, then there is no apparent reason to doubt the adequacy of the regression model, so linearity is implied.

5.4 Ultra-performance Liquid Chromatography

Using ultra-performance liquid chromatography (UPLC), it is now possible to take full advantage of chromatographic principles to run separations using shorter columns and higher flow rates for increased speed with superior resolution and sensitivity. Faster separations can lead to higher throughput and time savings when running of large samples is required. However, resolution in UPLC separation can reduce method development time from days, to hours or even minutes.

UPLC operates on the basics of HPLC with a column of submicron particle size (less than 2 µm) and high-pressure flow. Because of smaller particle size, higher back pressure is generated. Therefore, it becomes necessary to have a chromatography system that can operate at a pressure of 10,000 psi. As compared to particle size of 5 or 3 µm stationary phase, higher resolution and sensitivity can be achieved with smaller particles (1–2 µm). The increased resolution leads to better separation, increased sensitivity and faster analysis (Wilson *et al.*, 2005; Rainville *et al.*, 2007). The need to handle elevated pressures in excess of 10,000 psi can be alleviated to some extent by the use of increased column temperature. An increase in temperature leads to a reduction in the viscosity of mobile-phase solvents. An increase in column temperature can result in selectivity and peak elution order. Increasing the column temperature also increases the optimum linear velocity required to run the column effectively.

5.5 Application of HPLC for Quantification of Bioactive Compounds in Some Important Indian Medicinal Plants

5.5.1 Kalmegh (*Andrographis paniculata*)

Andrographis paniculata (Burm.f.) wall.ex Nees (*Acanthaceae*), or Kalmegh, is an important medicinal plant finding uses in many Ayurvedic formulations. Diterpenoid compounds, andrographolides, are the main bioactive phytochemicals present in the leaves and herbage of *A. paniculata*. An HPLC method was developed and applied for the determination of andrographolide (AP1), neoandrographolide (AP2) and andrograpanin (AP3) in the different extracts of *A. paniculata*. Under optimized conditions, three standards (AP1, AP2 and AP3) were well resolved with relatively high sensitivity at mean retention times of 5.41, 8.53 and 16.67 min for AP1, AP2 and AP3, respectively, when absorption was measured at 210 nm. The optimized parameters are described below.

HPLC parameters
Mobile phase: acetonitrile (15%, solvent A) and methanol–water, 60:40 (85%, solvent B).
HPLC column: RP-18 column (250 × 4 mm, 5 µm Merck, India).
Flow rate: isocratic: 0.6 ml min^{-1}.
Detection wavelength: 210 nm.
The method validation parameters and chromatograms are described in Tables 5.1–5.4 and Figs 5.3 and 5.4, respectively (Kumar *et al.*, 2014).

5.5.2 *Terminalia* species

Terminalia is a large genus of deciduous trees of the flowering plant family *Combretaceae* comprising about 250 species distributed in the tropical regions of the world. About 16 species occur in India. The medicinal properties of *Terminalia* species depend on the species and its part; however, the presence of tannins in the fruit and bark is almost characteristic of this genus. The *Terminalia* species represents a rich source of phenolic acids, tannins, cyclic triterpenoids and flavonoids.

Table 5.1. Retention time, equation for calibration curve, linear range, LOD and LOQ of AP1, AP2 and AP3 using developed HPLC method.

Analyte	Retention time mean (RSD %)	Regression equation ($y = ax + b$)	R^2	Linear range (µg ml^{-1})	LOD (µg ml^{-1})	LOQ (µg ml^{-1})
AP1	5.41 (0.18)	$y = 72{,}700x - 34{,}900$	0.997	10–200	5	10
AP2	8.53 (0.28)	$y = 42{,}000x + 118{,}000$	0.999	10–100	5	10
AP3	16.67 (1.22)	$y = 78{,}400x + 50{,}600$	0.999	5–100	2	5

Note: RSD = relative standard deviation; LOD = limit of detection; LOQ = limit of quantification.

Table 5.2. Precision (RSD %) of the developed HPLC method at three different concentrations of AP1, AP2 and AP3.

Analyte	Concentration (µg ml⁻¹)	RSD (%) Interday	RSD (%) Intraday
AP1	20	0.89	6.94
	60	2.36	1.57
	100	2.14	6.36
AP2	10	1.16	0.51
	40	1.71	1.82
	80	3.82	0.89
AP3	10	5.40	3.16
	40	5.38	3.94
	80	2.71	1.61

Note: RSD = relative standard deviation.

Table 5.3. Recovery data for AP1, AP2 and AP3 ($n = 3$).

Analyte	Added concentration (µg ml⁻¹)	Recovery (%)
AP1	67	107
	84	103
	134	118
AP2	67	93
	84	82
	134	87
AP3	67	87
	84	93
	134	104

Table 5.4. System suitability and peak purity parameters for AP1, AP2 and AP3.

Parameter	AP1	AP2	AP3
USP plate count	5420	5126	6694
USP tailing	1.27	1.15	1.26
Capacity factor	2.71	4.68	9.86
Resolution	1.79	7.44	11.89
Selectivity	1.13	1.73	2.10
Purity angle	0.034	0.140	0.211
Purity threshold	0.272	0.284	0.256

Note: USP = United States Pharmacopeia.

Their exact chemical classes and levels vary in different *Terminalia* species. Gallic acid (GA), corilagin (CL), chebulagic acid (CB), ellagic acid (EA) and chebulinic acid (CN) are the major tannin-related constituents of *Terminalia* species. A rapid and simple HPLC method was

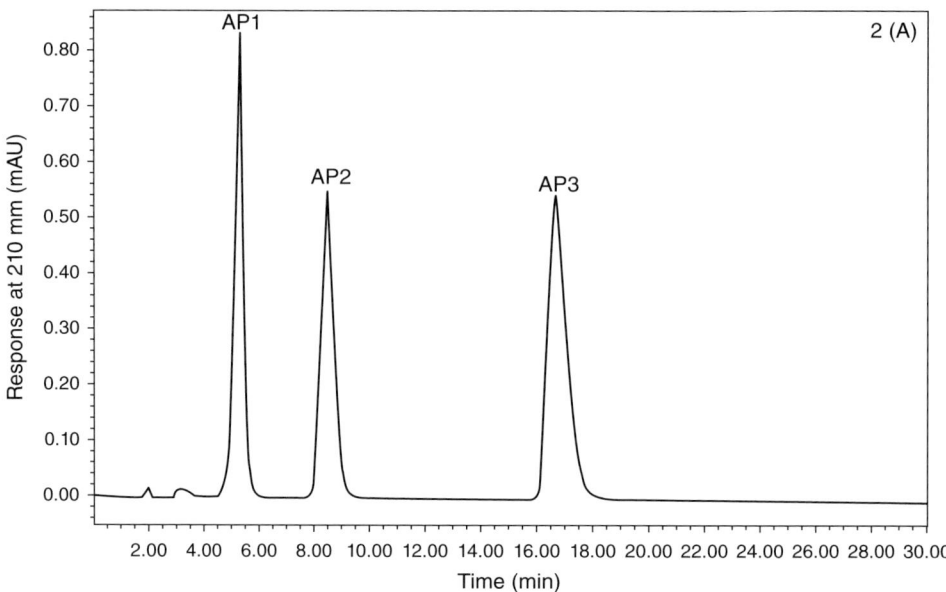

Fig. 5.3. HPLC chromatogram of standard mixture of AP1, AP2 and AP3.

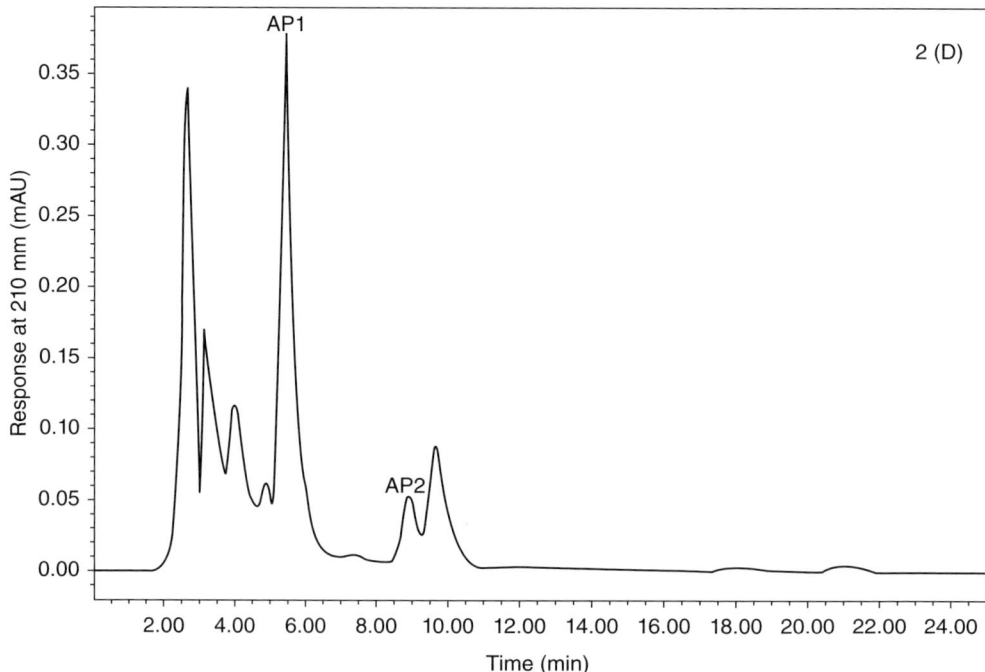

Fig. 5.4. HPLC chromatogram of supercritical fluid extract of leaves of *Andrographis paniculata*.

developed for the simultaneous identification and quantification of GA, CL, CB, EA and CN. The five tannin-related constituents were identified and quantified in the bark and fruit extracts of *Terminalia chebula*, *Terminalia bellirica*, *Terminalia arjuna* and *Terminalia catappa*.

HPLC parameters

Mobile phase: acetonitrile (A) and (0.05%) trifluoroacetic acid–water (B).
HPLC column: RP-18 column (150 × 4.6 mm, 5 μm, SunFire, waters, USA).
Flow rate: 1.0 ml min⁻¹.
Gradient: 0 min 90% B, 10 min 80% B, 15 min 72% B, 20 min 65% B, 25 min 50% B, then at 26 min, restoring the initial condition to 35 min.
Detection wavelength: 272 nm.
The method validation parameters and chromatograms are described in Tables 5.5–5.8 and Figs 5.5 and 5.6, respectively (Dhanani *et al.*, 2015).

5.5.3 *Vitex negundo* and *Vitex trifolia*

The genus *Vitex* belongs to the family *Lamiaceae*. It includes 80 genera and about 800 species. *Vitex negundo* is widely used in the indigenous system of medicine for its many medicinal properties, and in the past this plant has been studied extensively for its analgesic, anti-inflammatory, anticonvulsant and antioxidant activities. *Vitex trifolia* is known to possess antipyretic and antibacterial properties and is effective against asthma and allergic diseases. *V. trifolia* is known to produce a variety of diterpenoids and iridoids. Its leaves contain many bioactive phytochemicals such as flavonoids, sterols, diterpenoids, iridoids, etc. The leaves are considered useful as an external application for all rheumatic pain, sprains, etc. Roots are used to treat febrifuge, painful inflammations, coughs and fever. Flowers are used in the treatment of fever and fruit in amenorrhoea. Although all parts of *V. negundo* and *V. trifolia* are used in traditional medicines, the leaves are the most potent for medicinal uses. *p*-Hydroxybenzoic

Table 5.5. Retention time, equation for calibration curve, linear range, LOD, LOQ for GA, CL, CB, EA and CN using developed HPLC method.

Analyte	Retention time (min) mean	Regression equation ($y = ax + b$)	R^2	Linear range (μg ml⁻¹)	LOD (μg ml⁻¹)	LOQ (μg ml⁻¹)
GA	2.91	$y = 48,100x + 110,000$	0.999	2.5–75	1.0	2.5
CL	12.09	$y = 25,000x + 19,900$	0.997	1.0–75	0.5	1.0
CB	16.12	$y = 14,200x + 27,800$	0.998	2.5–75	1.0	2.5
EA	17.66	$y = 48,900x - 4,240$	0.992	1.0–75	0.5	1.0
CN	18.59	$y = 44,400x + 90,800$	0.994	2.5–100	1.0	2.5

Note: LOD = limit of detection; LOQ = limit of quantification; GA = gallic acid; CL = corilagin; CB = chebulagic acid; EA = ellagic acid; CN = chebulinic acid.

Table 5.6. Precision (RSD %) of the developed HPLC method at three different concentrations of GA, CL, CB, EA and CN using developed HPLC method.

Analyte	Concentration (µg ml^{-1})	RSD (%)	
		Interday	Intraday
GA	5	0.57	5.17
	25	2.82	5.98
	50	4.06	5.01
CL	5	7.99	2.09
	25	0.45	3.03
	50	3.32	4.40
CB	5	0.60	3.01
	25	5.86	3.66
	50	1.77	3.27
EA	5	6.61	5.63
	25	4.65	5.51
	50	0.07	3.0
CN	5	9.0	2.26
	25	0.39	4.21
	50	5.43	5.11

Note: RSD = relative standard deviation; GA = gallic acid; CL = corilagin; CB = chebulagic acid; EA = ellagic acid; CN = chebulinic acid.

Table 5.7 Recovery data for GA, CL, CB, EA and CN (*n* = 3).

Analyte	Added concentration (µg ml^{-1})	Recovery (%)
GA	5	100
	20	101
	40	103
CL	5	99
	20	103
	40	102
CB	5	98
	20	102
	40	103
EA	5	97
	20	96
	40	99
CN	5	98
	20	95
	40	100

Note: GA = gallic acid; CL = corilagin; CB = chebulagic acid; EA = ellagic acid; CN = chebulinic acid.

acid (PHBA), negundoside (NGN) and agnuside (AGN) are the active constituents of *V. negundo* and *V. trifolia* leaves. A HPLC method was developed to determine PHBA and AGN in different extracts of *V. negundo* and *V. trifolia* (Shah *et al.*, 2013).

Table 5.8. System suitability and peak purity parameters for GA, CL, CB, EA and CN.

Parameter	GA	CL	CB	EA	CN
USP plate count	8,209	12,278	37,071	51,523	88,393
USP tailing	1.17	0.77	0.74	1.10	0.74
Capacity factor	1.89	2.74	5.53	6.88	9.13
Resolution	–	28.02	14.27	5.73	4.50
Selectivity	2.10	5.29	1.49	1.12	1.07
Purity angle	0.147	0.201	0.357	0.334	0.327
Purity threshold	0.316	0.412	0.497	0.345	0.353

Note: GA = gallic acid; CL = corilagin; CB = chebulagic acid; EA = ellagic acid; CN = chebulinic acid; USP = United States Pharmacopeia.

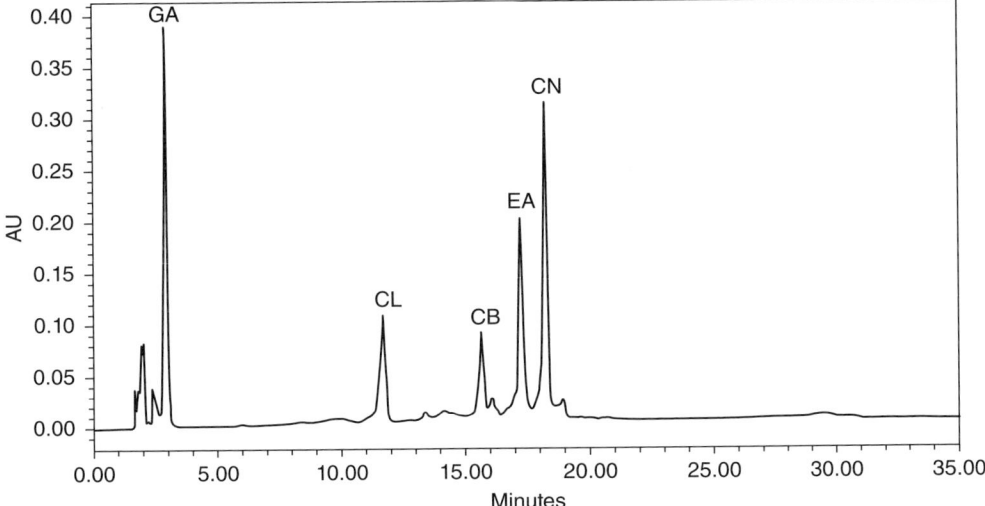

Fig. 5.5. HPLC chromatogram of standard mixture of gallic acid (GA), corilagin (CL), chebulagic acid (CB), ellagic acid (EA) and chebulinic acid (CN).

HPLC parameters
Solvent: acetonitrile (15%, A) and 0.5% *O*-phosphoric acid–water (85%, B).
HPLC column: RP-18 column (250 × 4.6 mm, 5 µm, Merck, India).
Flow rate: isocratic, 1.0 ml min^{-1}.
Detection wavelength: 254 nm.
The method validation parameters and chromatograms are described in Tables 5.9 and 5.10 and Figs 5.7 and 5.8, respectively.

5.5.4 Mamejo (*Enicostemma axillare*)

Enicostemma axillare Lam. Raynal (Syn. *E. littorale* Blume), a perennial glabrous herb 10–50 cm high, is distributed throughout India up to an altitude of 457 m, from Punjab and the Gangetic plain to Ceylon, and is found

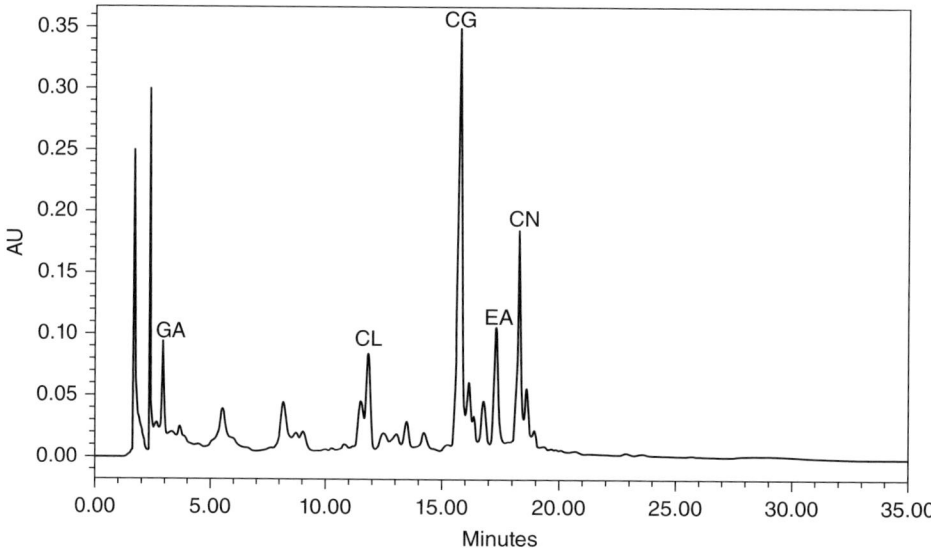

Fig. 5.6. HPLC chromatogram of fruit pericarp extract of *Terminalia chebula*.

Table 5.9. Intraday and interday precision data of developed HPLC method.

Standard (µg ml⁻¹)	Intraday (RSD %)	Interday (RSD %)
PHBA		
5	0.28	0.30
40	0.45	2.27
100	0.43	1.74
AGN		
25	0.07	0.41
100	0.03	0.44
250	0.05	0.26

Note: RSD = relative standard deviation; PHBA = *p*-hydroxybenzoic acid; AGN = agnuside.

Table 5.10. Analytical recovery of PHBA and AGN by developed HPLC method (*n* = 3).

Standard	Added concentration (µg ml⁻¹)	Measured concentration (µg ml⁻¹)	Recovery (%)	RSD (%)	Mean recovery (%)
PHBA	20	21.88	109.41	2.43	
	40	40.16	100.41	1.90	106.11
	80	86.91	108.53	0.54	
AGN	50	42.65	85.30	0.29	
	200	196.46	98.23	0.54	93.07
	400	382.74	95.68	0.18	

Note: PHBA = *p*-hydroxybenzoic acid; AGN = agnuside; RSD = relative standard deviation.

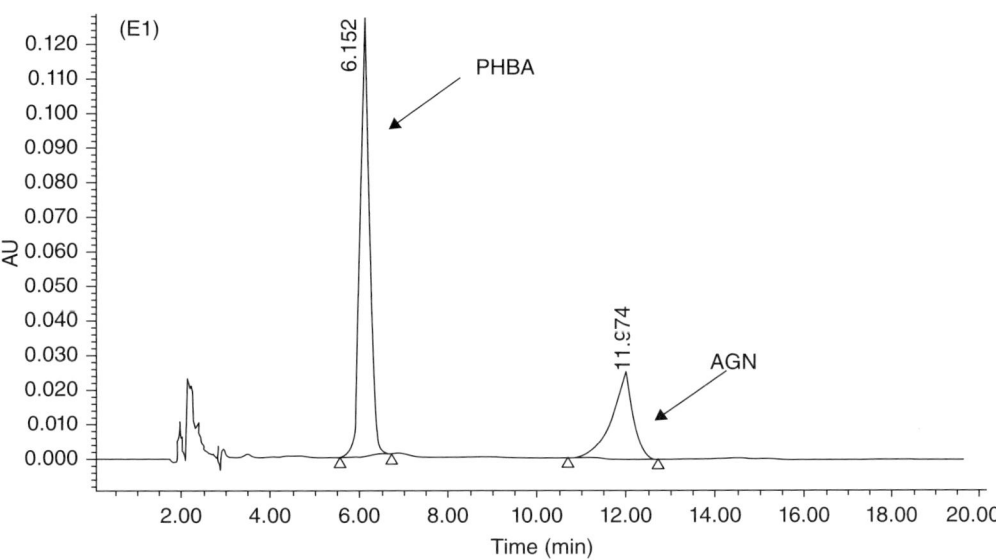

Fig 5.7. HPLC chromatogram of standard mixture of PHBA and AGN.

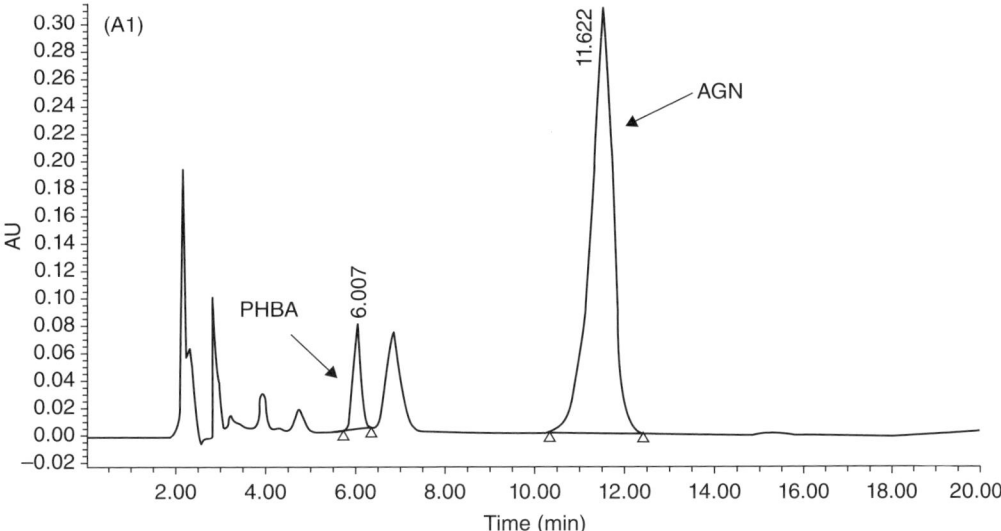

Fig. 5.8. HPLC chromatogram methanol extract of *Vitex negundo* leaves.

more frequently near the sea. The plant is bitter and used for curing fever, snakebite and vata diseases, and as a tonic, antihelmintic, stomachic, laxative and hypolipidemic. The powdered plant is given with honey as a blood purifier and to help cure dropsy, rheumatism, abdominal ulcers, hernia, swellings, itches and insect poisoning. Extracts of *E. axillare* have been reported to have antidiabetic, antioxidant, hepatoprotective and anti-inflammatory activities. The aerial part of *E. axillare* is known to possess potent medicinal value and is used in the treatment of many diseases in India. Swertiamarin, a seco-iridoid, has also been identified as a major compound in its aerial parts and found to be a bioactive constituent. A simple, rapid and precise HPLC method for the estimation of a bioactive swertiamarin in the aerial parts was developed and validated (Rana *et al.*, 2012). Quantification of swertiamarin at 238 nm wavelengths was found to be better as compared to 227 nm and 254 nm wavelengths.

HPLC parameters
Mobile phase: methanol–water (1:1).
HPLC column: RP-18 column (250 × 4.6 mm, 5 µm, Lichro CART®, Merck, Germany).
Flow rate: flow gradient started with solvent A (flow rate of 0.5 ml min^{-1}). At 4.9 min, the flow rate was increased to 0.8 ml min^{-1}. The flow rate was then increased to 1.0 ml min^{-1} at 5.1 min. The flow rate was kept constant up to 5.4 min and then finally decreased to 0.5 ml min^{-1}, restoring the initial conditions at 8 min.
Detection wavelength: 238 nm.
The method validation parameters and chromatograms are described in Tables 5.11 and 5.12 and Figs 5.9 and 5.10, respectively.

Table 5.11. Method validation data for determination of swertiamarin by HPLC.

Parameter	Value
Mean retention time	5.26 min (RSD 0.5%)
Linear range	4–80 µg ml^{-1}
Regression equation	$y = 18,000x - 1,380$
Regression coefficient (R^2)	0.998
Limit of detection (LOD)	4 µg ml^{-1}
Limit of quantification (LOQ)	6 µg ml^{-1}

Note: RSD = relative standard deviation; LOD = limit of detection; LOQ = limit of quantification.

Table 5.12. Precision (RSD) of the developed HPLC method at three different concentrations of swertiamarin.

Concentration (µg ml^{-1})	RSD (%) Intraday	Interday
Low (6)	0.12	2.68
Medium (20)	0.10	2.71
High (60)	1.59	2.88

Note: RSD = relative standard deviation.

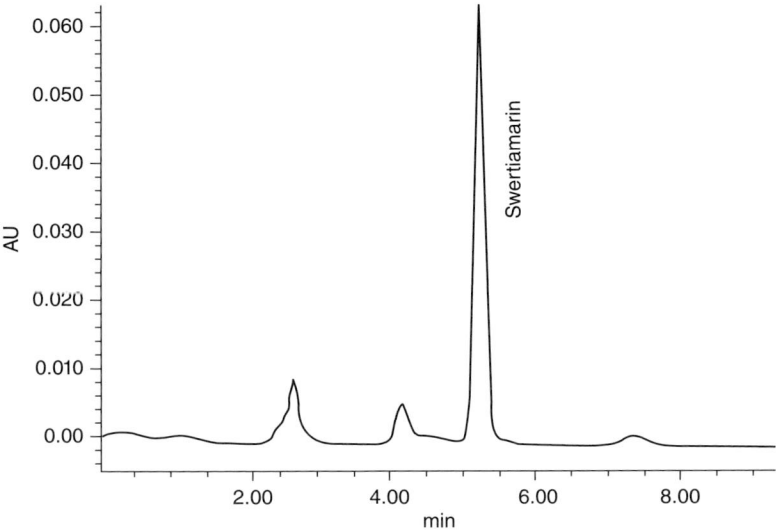

Fig. 5.9. HPLC chromatogram of water extract of aerial parts of *Enicostemma exillare*.

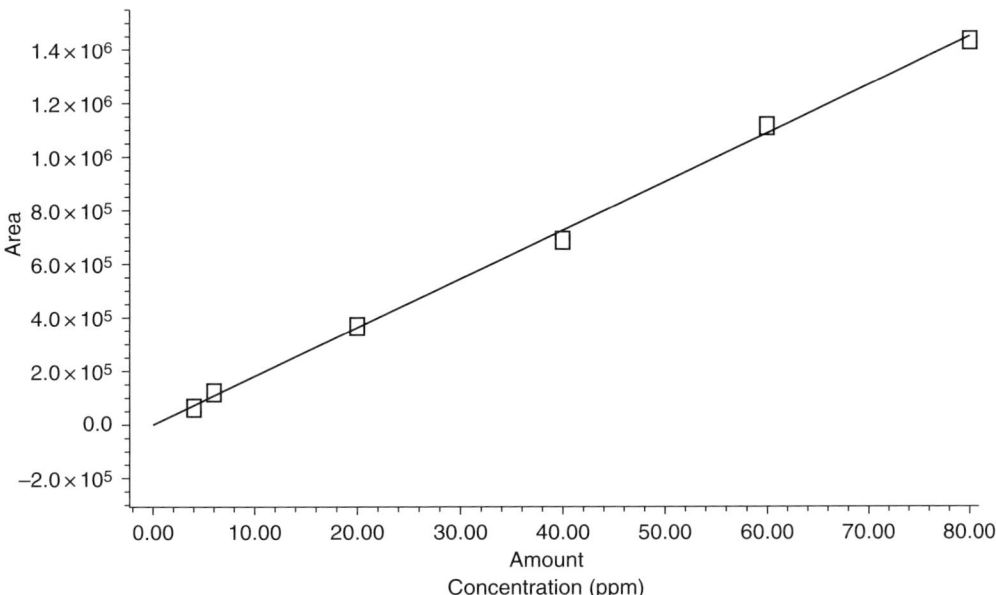

Fig. 5.10. Calibration curve of different concentrations of swertiamarin.

5.6 Conclusion

Major components could be present in several closely related species. The value of the HPLC method developed increases with the estimation of all the characteristic compounds present in the extracts.

References

Dey, A.K., Datta, S. and Mulherjee, A. (2012) Quantitative analysis of glycyrrhzic acid from a polyherbal preparation using liquid chromatographic technique. *Journal of Advanced Pharmaceutical Technology & Research* 3, 210–215.

Dhanani, T., Shah, S. and Kumar, S. (2015) A validated high-performance liquid chromatography method for determination of tannin-related marker constituents gallic acid, corilagin, chebulagic acid, ellagic acid and chebulinic acid in four *Teminalia* species from India. *Journal of Chromatographic Science* 53, 625–632.

Dixon, R.W. and Peterson, D.S. (2002) Development and testing of a detection method for liquid chromatography based on aerosol charging. *Analytical Chemistry* 74, 2930–2937.

Dongre, V.G., Karmuse, P.P., Ghugare, P.D., Gupta, M., Nerukar, B., *et al.* (2007) Characterization and determination of impurities in piperaquine phosphate by HPLC and LC/MS/MS. *Journal of Pharmaceutical and Biomedical Analysis* 43, 186–195.

Engelhardt, H. and Siffrin, C. (1997) Means of validation of HPLC systems and components. *Chromatographia* 47, 36–43.

Ford, D.L. and Kennard, W. (1966) Vaporisation analyzer. *Journal of Oil Colour Chemists' Association* 49, 299–313.

Foster, M.D. and Synovec, R.E. (1996) Reversed phase liquid chromatography of organic hydrocarbons with water as the mobile phase. *Analytical Chemistry* 68, 2838–2844.

Ganzera, M., Gampenrieder, J., Pawar, R.S., Khan, I.A. and Stuppner, H. (2004) Separation of the major triterpenoid saponins in *Bacopa monnieri* by high-performance liquid chromatography. *Analytica Chimica Acta* 516, 149–154.

Guillarme, D., Rudaz, S., Schelling, C., Dreux, M. and Veuthey, J.L. (2008) Micro liquid chromatography coupled with evaporative light scattering detector at ambient and high temperature: optimization of the nebulization cell geometry. *Journal of Chromatography A* 1192, 103–112.

Guo, C.J., Cao, G.H., Sofic, E. and Prior, R.L. (1997) High-performance liquid chromatography coupled with coulometric array detection of electroactive components in fruit and vegetables: relationship to oxygen radical absorbance capacity. *Journal of Agricultural and Food Chemistry* 45, 1787–1796.

International Conference on Harmonisation (ICH) (2005) Q2 9 (R1): Technical requirements for registration of pharmaceuticals for human use, validation of analytical procedures: text and methodology. In: *Requirements for Registration of Pharmaceuticals for Human Use*. ICH, Geneva, Switzerland, pp. 1–13.

Jaimez, J., Fente, C.A., Vazquez, B.I., Franco, C.M., Cepeda, A., *et al.* (2000) Application of the assay of aflatoxins by liquid chromatography with fluorescence detection in food analysis. *Journal of Chromatography A* 882, 1–10.

Kristl, J., Veber, M., Krajnicic, B., Oresnik, K. and Slekovec, M. (2005) Determination of jasmonic acid in *Lemna minor* (L.) by liquid chromatography with fluorescence detection. *Analytical and Bioanalytical Chemistry* 383, 886–893.

Kumar, S., Dhanani, T. and Shah, S. (2014) Extraction of three bioactive diterpenoids from *Andrographis paniculata*: effect of the extraction techniques on extract composition and quantification of three andrographolides using high-performance liquid chromatography. *Journal of Chromatographic Science* 52, 1043–1050.

LaCourse, W.R. and Modi, S.J. (2005) Microelectrode application of pulsed electrochemical detection. *Electroanalysis* 17, 1141–1152.

Li, F.M., Zhang, C.H., Guo, X.J. and Feng, W.Y. (2003) Chemiluminesscence detection in HPLC and CE for pharmaceutical and biomedical analysis. *Biomedical Chromatography* 17, 96–105.

Megoulas, N.C. and Koupparis, M.A. (2005) Twenty years of evaporative light scattering detection. *Critical Reviews in Analytical Chemistry* 35, 301–316.

Neue, U.D. and Mazzeo, J.R. (2001) A theoretical study of the optimization of gradients at elevated temperature. *Journal of Separation Science* 24, 921–929.

Nguyen, D.T.T., Guillarme, D., Rudaz, S. and Veuthey, J.L. (2006) Fast analysis in liquid chromatography using small size and high pressure. *Journal of Separation Science* 29, 1836–1848.

Ohba, Y., Kuroda, N. and Nakashima, K. (2002) Liquid chromatography of fatty acids with chemiluminescence detection. *Analyica Chimica Acta* 465, 101–109.

Ozkan, S.A. (2007) LC with electrochemical detection. Recent application to pharmaceuticals and biological fluids. *Chromatographia* 66, S3–S13.

Rainville, P.D., Stumf, C.L., Shockcor, J.P., Plumb, R.S. and Nicholson, J.K. (2007) Novel application of reversed phase UPLC-oa TOF-MS for lipid analysis in complex biological mixtures: a novel tool for lipidomics. *Journal of Proteome Research* 6, 552–558.

Rana, V.S., Dhanani, T. and Kumar, S. (2012) Improved and rapid HPLC-PDA method for identification and quantification of swertiamarin in the aerial parts of *Enicostemma axlllare*. *Malaysian Journal of Pharmaceutical Sciences* 10, 1–10.

Shah, S., Dhanani, T. and Kumar, S. (2013) Validated HPLC method for identification and quantification of *p*-hydroxybenzoic acid and agnuside in *Vitex negundo* and *Vitex trifolia*. *Journal of Pharmaceutical Analysis* 3, 500–508.

Sun, S.W. and Su, H.T. (2000) Validated HPLC method for determination of the sennosides A and B in senna tablets. *Journal of Pharmaceutical and Biomedical Analysis* 29, 881–894.

Swartz, M.E. (2005) UPLC: an introduction and review. *Journal of Liquid Chromatography and Related Technologies* 28, 1253–1263.

Synder, R.L., Kirkland, J. and Glajch, L. (1997) *Practical HPLC Method Development*. Wiley-VCH, New York.

Vervoort, N., Daemen, D. and Torok, G. (2008) Performance evaluation of evaporative light scattering detection and charged aerosol detection in reversed phase liquid chromatography. *Journal of Chromatography A* 1189, 92–100.

Willard, H.H., Meritt, L.L. Jr, Dean, J.A. and Settle, F.A. Jr (1986) *Instrumental Methods of Analysis*. CBS Publishers & Distributors, New Delhi.

Wilson, I.D., Nicholson, J.K., Castro-Perez, J., Granger, J.H., John, K.A., *et al.* (2005) High-resolution 'ultra performance' liquid chromatography coupled to oa-TOF mass spectrometry as a tool for differential metabolic pathway profiling in functional genomic studies. *Journal of Proteome Research* 4, 591–598.

Wolfender, J.L. (2009) HPLC in natural product analysis: the detection issue. *Planta Medica* 75, 719–734.

Zhang, Q.L. and Cui, H. (2005) Simultaneous determination of quercetin, kaempferol and isorhamnetin in phytopharmaceuticals of *Hippophae rhamnoids* by high-performance liquid chromatography with chemiluminescence detection. *Journal of Separation Science* 28, 1171–1178.

6 Tandem Techniques

6.1 Introduction

Phytochemicals are formed as a result of biochemical transformations in plants. They are present often in low or very low concentrations in a complex matrix. The analysis of individual natural products in complex crude extracts requires efficient separation methods prior to their detection. According to the type of study (quantification, standardization, fingerprinting, screening, etc.), very sensitive and selective methods may be needed.

Hyphenated techniques combining different separation and detection methods were first introduced by Tomas Hirschfeld about three decades ago (Wilson and Brinkman, 2003). Hyphenated techniques allow a rapid structural determination of known plant constituents with a minute amount of plant material (Wolfender *et al.*, 2000). With such a combined approach, the time-consuming isolation of common natural products is avoided and an efficient targeted isolation of compounds presenting interesting spectroscopic or biological features is performed. To discover new bioactive products, the dereliction of crude extracts performed prior to isolation work is of crucial importance in order to avoid the tedious isolation of known constituents. Hyphenated analytical methods have been fully integrated into the isolation process and are used for chemical screening of crude plant extracts.

Chromatographic fingerprinting has gained more and more attention in recent times. It has been widely introduced and accepted by the World Health Organization (WHO), the Food and Drug Administration (FDA), the European Medicines Agency (EMA), the German Commission, the British Herbal Medicine Association, the Indian Drug Manufacturer's Association and many other official organizations as a strategy to assess the quality of herbal products. A chromatographic fingerprint is a chromatogram that represents the chemical characteristics of the constituent analyte, and it

 © Satyanshu Kumar 2016. *Analytical Techniques for Natural Product Research* (S. Kumar)

has the potential to identify the genuine raw material as well as determining the identity, authenticity and batch-to-batch consistency of herbal medicines. The authentication and identification of a drug and its product can be performed accurately using chromatographic fingerprints, even if batches or concentration vary among samples.

Combining a chromatographic separation system online with a spectrometer detector in order to obtain structural information from the analytes present in a sample has become the most important approach for the identification and confirmation of the identity of target and unknown chemical compounds. Hyphenated techniques such as liquid chromatography–ultraviolet photodiode array detection, liquid chromatography–mass spectrometry, liquid chromatography–tandem mass spectrometry and liquid chromatography–nuclear magnetic resonance are being used more and more extensively. The data obtained from such hyphenated techniques are called two-way data, one way for the chromatogram and the other way for the spectrum, which could provide much more information than the classic one-way chromatogram, as sometimes a single chromatogram turns out to be inadequate for complex herbal products.

Recent advances in hyphenated analytical techniques have widened their application remarkably to natural products. Hyphenated techniques are being applied to the pre-isolation and isolation of natural products, dereplication, online partial identification of compounds, chemotaxonomic studies, chemical fingerprinting, quality control and metabolic studies. These techniques offer simultaneous separation and structural information on compounds present in complex mixtures (Exarchou *et al.*, 2006). A method where a separation technique is coupled with one or more online spectroscopic detection techniques such as ultraviolet–visible (UV-vis), infrared (IR), mass (MS) or nuclear magnetic resonance (NMR) spectroscopy is known as a hyphenated technique (Sarkar and Nahar, 2012). The following are the most commonly used hyphenated techniques in natural products chemistry:

1. Gas chromatography–mass spectrometry (GC-MS).
2. Liquid chromatography–mass spectrometry (LC-MS).
3. Liquid chromatography–nuclear magnetic resonance spectrometry (LC-NMR).
4. Capillary electrophoresis–mass spectrometry (CE-MS).
5. Supercritical fluid extraction–supercritical fluid chromatography (SFE-SFC).

Hyphenation is not necessarily limited to two techniques only. The coupling between the separation and detection techniques can involve more than one separation or detection technique, resulting in multiple hyphenated techniques such as LC-PDA (photodiode array detection)-MS, LC-MS-MS, LC-NMR-MS and LC-PDA-NMR (Sarkar and Nahar, 2012). Combining a chromatographic system online with a spectroscopic detector has become the most important approach for the identification and or confirmation of the identity of the target and unknown chemical compounds. The most preferred approach is the combination of column liquid chromatography or capillary gas chromatography with a mass spectrometer

(LC-MS and GC-MS, respectively). In quite a number of cases, additional or complementary information is also essentially required. This information can be obtained by Fourier transform infrared, diode array UV–visible absorbance or fluorescence emission or nuclear magnetic resonance spectrometry (Exarchou *et al.*, 2006). The use of all these hyphenated techniques allows the rapid structural determination of known plant constituents with a minute amount of plant material. By following this approach, the time-consuming isolation of common natural products is avoided and an efficient targeted isolation of compounds presenting interesting spectroscopic or biological features can be performed. The selection of a hyphenated method is done on the basis of analyte nature, possible combination of various methods and determined sensitivity and apparatus availability. Albert (1995) described direct online coupling between high-performance liquid chromatography (HPLC), supercritical fluid chromatography (SFC) and capillary electrophoresis (CE) and proton high field NMR spectroscopy. It has been reported that the resolution of the ^{1}H NMR spectra obtained in HPLC-NMR, SFC-NMR and SFE-NMR coupling under continuous flow conditions was similar to conventionally recorded NMR spectra. In CE-NMR coupling, signal line widths were degraded, but the resolution of CE-NMR spectra was improving continuously.

6.2   Gas Chromatography–Mass Spectrometry (GC-MS)

Hyphenated methods were first introduced through coupling GC with different detectors in the following systems: gas chromatography–atomic absorption spectrometry (GC-AAS); gas chromatography–atomic emission spectrometry (GC-AES); gas chromatography–mass spectrometry (GC-MS); or gas chromatography–inductively coupled plasma mass spectrometry–time of flight (GC-ICP-MS-TOF). Mass spectrometry is the most popular detection method as it offers information on quantitative and qualitative sample compositions and helps to determine sample composition. Mass spectrometry is the most sensitive and selective method for molecular analysis and provides information on the molecular weight as well as the structure of the molecule. Combining chromatography with mass spectrometry provides the advantage of both chromatography as a separation method and mass spectrometry as an identification method. With the combination of the retention time and accurate molecular mass, fast quantification is possible even at very low concentrations. However, the main difficulties in coupling the MS detector with chromatographic methods arise from the necessity to maintain very low pressure in the spectrometers. In mass spectrometry, there is a range of methods to ionize compounds and then separate the ions. Electron impact (EI) and electron capture ionization (ECI) are the common methods of ionization used in conjugation with gas chromatography.

Instrumentation for GC-MS consists mainly of an injector port, column and mass spectrometric detector. Carrier gas (argon/helium/nitrogen) drives the sample to the column. Two types of columns, capillary and packed, are generally used.

Compounds that are adequately volatile and stable in high temperature can be analysed easily by GC-MS. Sometimes, the derivatization of compounds to make them volatile is carried out prior to GC-MS analysis. Sample injected into the injector port of a gas chromatography system is separated in the column and detected in the mass spectrometric detector. Interpretation of fragments with different relative abundance in mass spectral data provides information about the structure of the compound when compared with the reference spectra library. EI is configured primarily to select positive ions, whereas ECI is usually configured for negative ions. However, various other types of detection system, such as TOF, are also used nowadays. EI provides reproducible mass spectra with structural information, which allows library searching and is particularly useful for routine analysis.

GC-MS is used in a wide range of applications, such as flavour and fragrance analysis, pesticide analysis and metabolite analysis. It is considered as the method of choice for the detection of volatile compounds due to its high sensitivity over other analytical techniques like LC-MS. Also, with the selection of a suitable column, a wide range of compounds can be analysed. The results obtained with GC-MS are more confirmatory to GC. Furthermore, as a powerful tool, it is being used increasingly in biomarker discovery.

GC-MS has been demonstrated to be an important technique for the analysis non-polar and volatile natural products such as essential oils comprising mono and sesquiterpenes. Furthermore, with the capillary column, GC-MS has a very good separation ability, which produces a high-quality fingerprint. Also, coupling with the mass spectrometer and the corresponding mass spectral database, qualitative and relatively quantitative composition information could be obtained, which is extremely useful for further research to elucidate the relationship between chemical constituents and its pharmacology. Chen *et al.* (1987) reported the application of this hyphenated technique in determining about 130 volatile constituents in several Chinese medicinal herbs. Essential oils from *Pestacia atlantica* var. *mutica*, composed mainly of monoterpenes, were characterized using GC-MS (Delazar *et al.*, 2004).

6.3   High-performance Liquid Chromatography–Photodiode Array Detection (HPLC-PDA)

Chromatography had been known as a separation method since the beginning of the 20th century (Tswett, 1906). Its rapid development occurred about 50 years later. Nowadays, chromatographic methods are the most popular instrumental analytical methods in analytical chemistry, as they allow quick separation and determination of substances in a complex matrix (Michalski *et al.*, 2013).

HPLC is a very powerful and versatile chromatographic technique for the separation of natural products in a complex matrix without the need for complex sample preparation (Kingston, 1979). HPLC has developed greatly over the years in terms of convenience, speed, choice of column stationary phases, high sensitivity and applicability to a broad variety of sample matrices, plus the ability to hyphenate the chromatographic method to spectroscopic detectors (Natishan, 2004). The development of columns with different phase chemistry, especially reversed phase, enabled the separation of almost any types of natural products. Photodiode array detection (PDA) provides UV spectra directly online, and is particularly useful for the detection of natural products with characteristic chromophores (Larsen and Hansen, 2008). As all wavelengths are stored during analysis, multiple wavelengths can be monitored at the same time to detect different classes of compounds. HPLC-PDA has become a common technique in most analytical laboratories worldwide as it is easy to learn and use. Over the past decades, HPLC has received the most extensive application in the analysis of natural products. Optimal separation conditions for HPLC depend on factors such as mobile phase compositions, their pH adjustment, pump pressure, etc. The advantages of HPLC lie in its versatility in the analysis of chemical compounds. The qualitative analysis of complex samples of natural products is much easier now because of the additional UV spectral information obtained from a photodiode array detector. However, many chemical compounds in natural products are non-chromophoric compounds; therefore, a single-wavelength UV detector is not capable of fulfilling this requirement. Evaporative light scattering detection (ELSD) is an excellent detection method for non-chromophoric compounds (Niemi *et al.*, 1997). The response of ELSD depends only on the size, shape and number of eluate particles rather than on the analyte structure or the chromophore of the analyte. Checking the peak purity as well as comparing it with the available reference standard spectrum of the known compound to the one in the investigated sample is much easier than before. However, structural elucidation is not possible by a simple HPLC. In multiple hyphenated systems with mass spectrometric detection, HPLC- PDA systems are commonly used.

HPLC-PDA methods can be used to compare the fingerprint profiles of closely related species, or the same species from different locations. For *Herba Oldenlandiae*, a plant of the Chinese pharmacopoeia, which can be adulterated by different *Oldenlandia* species (*Oldenlandia corymbosa* (L.) Lam and *Oldenlandia tenelliflora* Bl.), a comparative analysis of chemical components was obtained by calculating the correlation coefficient of the entire UV chromatograms to appraise similarity (Liang *et al.*, 2007). The chemical profile of *Oldenlandia diffusa* was quite different in samples collected from different habitats. In species delimitation, where it is not possible using morphological characteristics, the HPLC fingerprinting method coupled with multivariate analysis can be used for identification and classification purposes.

Fingerprint-based similarity for a chemotaxonomic study was performed for different *Taxus* species by the application of hierarchical clustering analysis (HCA) and principal component analysis (PCA) to a set of HPLC-UV profiles (Ge *et al.*, 2008). Based on the PCA loadings, 12 chemical constituents were selected as chemotaxonomic markers. Based on the results, eight species could be divided into six well-supported groups, and most samples could be assigned to the correct species.

The structural information gained from this detector is very limited. This is also true of other, more specialized detectors such as fluorescence or evaporative light scattering detectors.

6.4   Liquid Chromatography–Mass Spectrometry (LC-MS)

Phytochemical analysis requires the isolation of plant extract ingredients in adequate quantities for spectral and biological analysis. The rapid detection of biologically active natural products plays a strategic role in the phytochemical investigation of crude plant extracts. For efficient screening of the extracts, quantitative as well as quantitative analysis using HPLC and biological assays are used. Combining chromatographic and spectroscopic techniques is a well-known strategy, and a technique utilizing atmospheric pressure ionization, namely atmospheric chemical ionization (APCI) and electrospray ionization (ESI), has led to the development of coupling liquid chromatography (HPLC) with mass spectrometry (Herderich *et al*, 1997).

Hyphenated techniques such as HPLC coupled with a UV photodiode array detector (HPLC-DAD-UV) and a mass spectrometer provide online much useful structural information on metabolites prior to isolation. Chromatographic methods are used mainly for separation, and spectroscopic techniques enable detection (Ellis and Roberts, 1997; Shalliker, 2011). Hyphenated methods also include the coupling of several chromatographic methods (Purcaro *et al.*, 2012). HPLC-MS, also known as LC-MS, is the most popular technique for metabolic profiling and fingerprinting as it provides very important information about the molecular weights of compounds separated by liquid chromatography. This technique is becoming of increasing importance, mainly because of the enhanced performance now attainable in chromatography (nano HPLC, ultra performance liquid chromatography (UPLC)) and mass spectrometry (high resolution and high throughput). Furthermore, LC-MS and LC-NMR techniques have gradually limited the need to isolate individual compounds in pure form prior to the identification step, which has helped enormously in avoiding the isolation of unnecessary compounds. LC-MS has been more popular than LC-NMR because of its greater sensitivity and lower cost (Sarkar and Nahar, 2012).

In the past 20 years, the online combination of LC and MS has become a robust and routinely applicable tool. The separated analytes emerging from the column are identified on the basis of mass spectral data.

The LC-MS system consists of a liquid chromatography system with mass spectrometer. It is sensitive and can be considered both a specific and a universal detection method.

The multiple reaction-monitoring mode (MRM) offers good sensitivity because it decreases noise levels significantly and accordingly increases the response of analytes. The target compounds are identified on the basis of the mass spectra of the parent and the product obtained from the parent (Figs 6.1a and b and 6.2a and b).

The application of selected reaction monitoring (SRM) or multiple reaction monitoring in tandem mass spectrometry (MS/MS) allows the selection and fragmentation of a parent ion, followed by selective monitoring of the product ion. These transitions enable very sensitive detection and

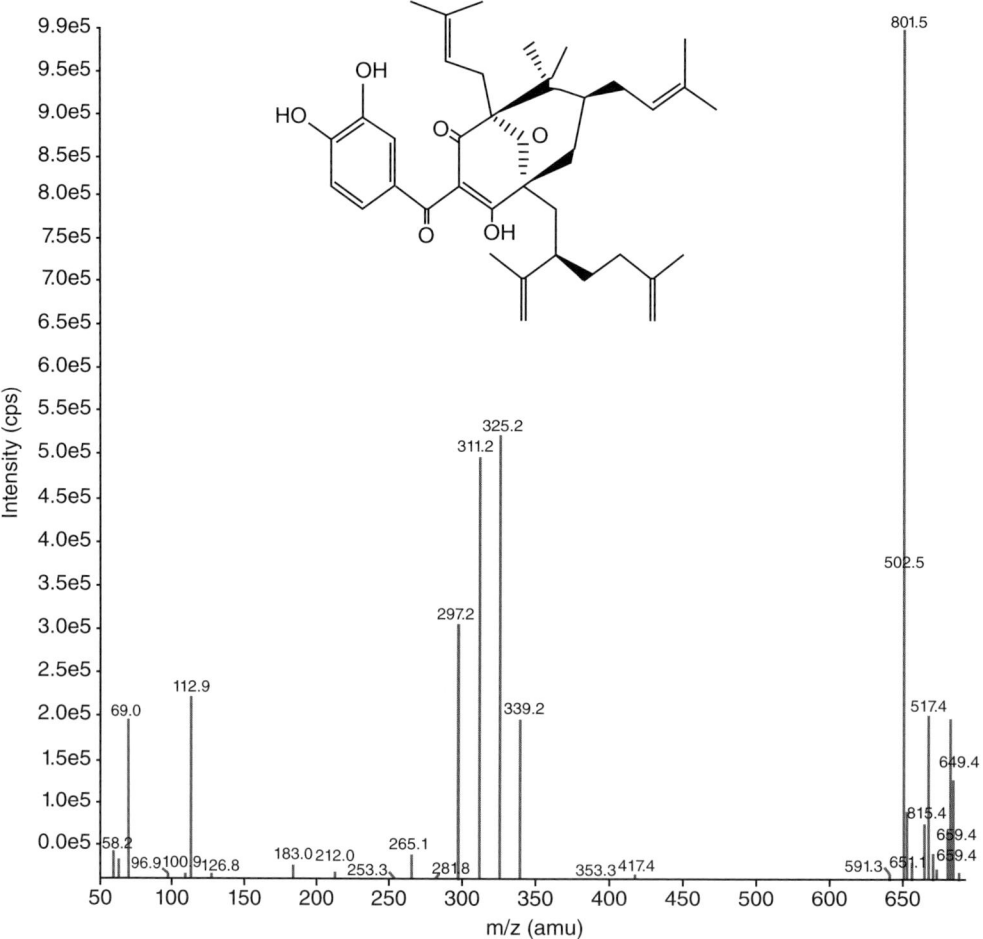

Fig. 6.1a. Primary mass spectrum (Q1 MS) of xanthochymol ($C_{38}H_{50}O_6$) in negative ionization mode. (From Chattopadhyay and Kumar, 2006.)

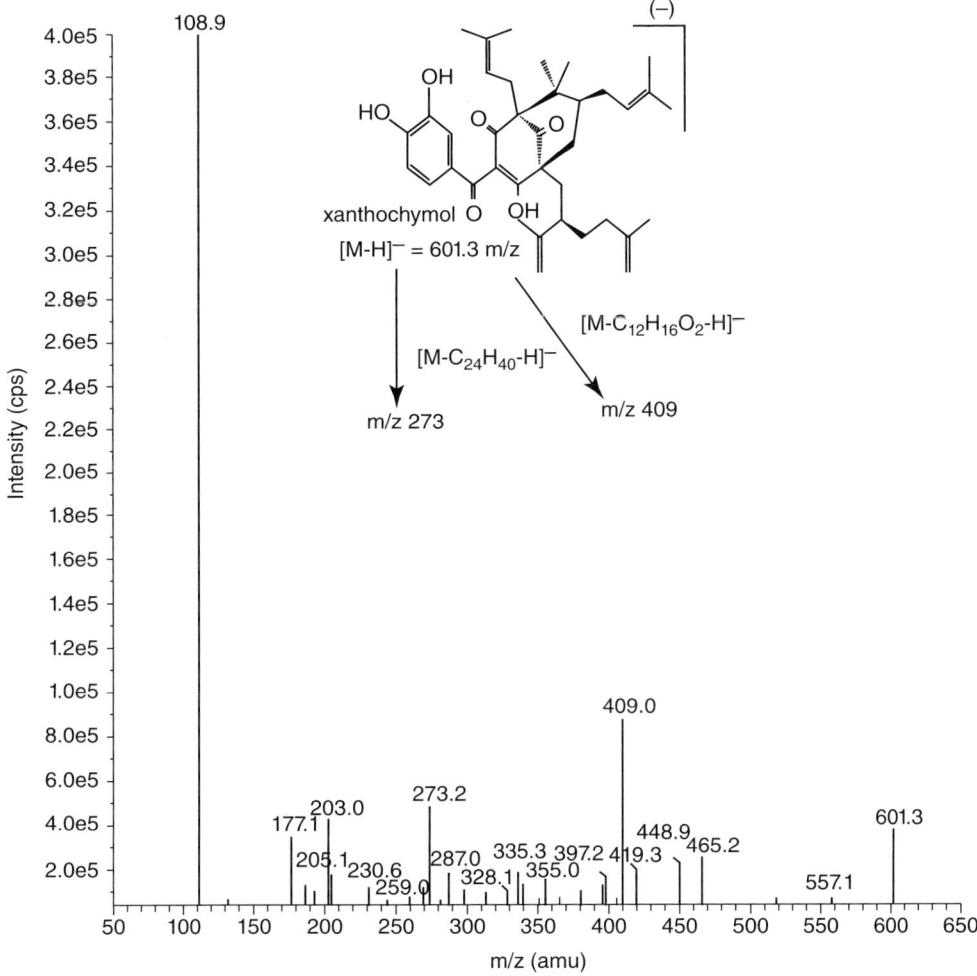

Fig. 6.1b. Product ion mass spectrum (Q2 MS) of xanthochymol ($C_{38}H_{50}O_6$) in negative ionization mode. (From Chattopadhyay and Kumar, 2006.)

quantification of compounds because there is virtually no background. Separated analyte ions leave the chromatographic column under relatively high pressure. Analyte ions are separated in the spectrometer on the basis of the mass-to-charge ratio, and the collected data are obtained in the form of a spectrum. The ionization chamber is the fundamental part of the spectrometer. It can be coupled with any available mass analyser, such as quadruple, ion trap, TOF, sector field and Fourier transform into cyclotron resonance or hybrid couplings, such as quadruple with the time of flight.

ESI, APCI or atmospheric pressure photochemical ionization (APPI) are the different ionization methods used in mass spectrometers. ESI and APCI are the most popular methods as they solve most analytical problems related to both big and small molecules. However, APCI has been

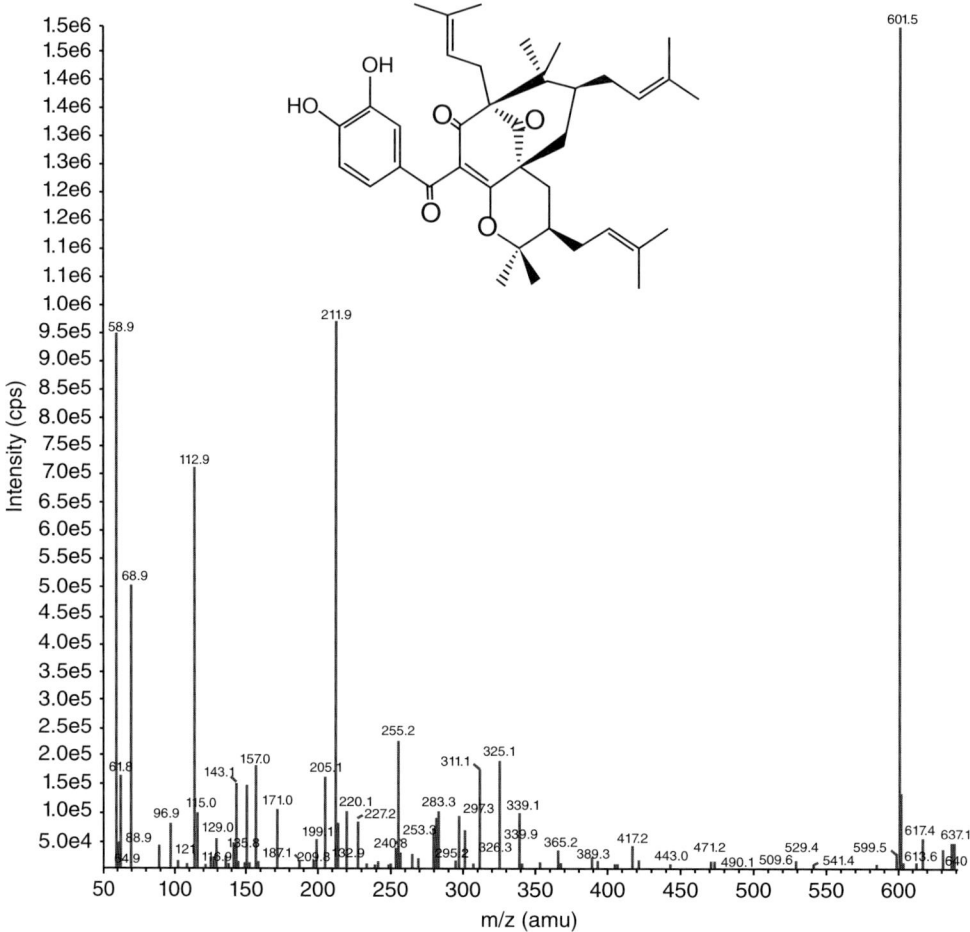

Fig. 6.2a. Primary mass spectrum (Q1 MS) of isoxanthochymol ($C_{38}H_{50}O_6$) in negative ionization mode. (From Chattopadhyay and Kumar, 2006.)

considered mainly applicable to the assay of weakly polar and stable compounds. APCI permits higher HPLC flow rates.

ESI-MS is used to determine compounds in biological materials; that is, nucleic acids, amino acids, peptides, proteins and their complexes with metals and metalloids. APPI is a modern solution and is employed chiefly to analyse small particles, including the non-polar particles. All the techniques discussed belong to the so-called soft ionization methods.

Selected ion monitoring (SIM) and scan modes are used for detection in the mass spectrometer. Quadruple analysers are used most often. They are comparatively inexpensive and are easy to operate. They provide quick data transfer and fast separation of ions on the basis of the mass-to-signal ratio. The resolution increase leads to an inevitable drop in method sensitivity. Importantly, the quadruple spectrometers now used

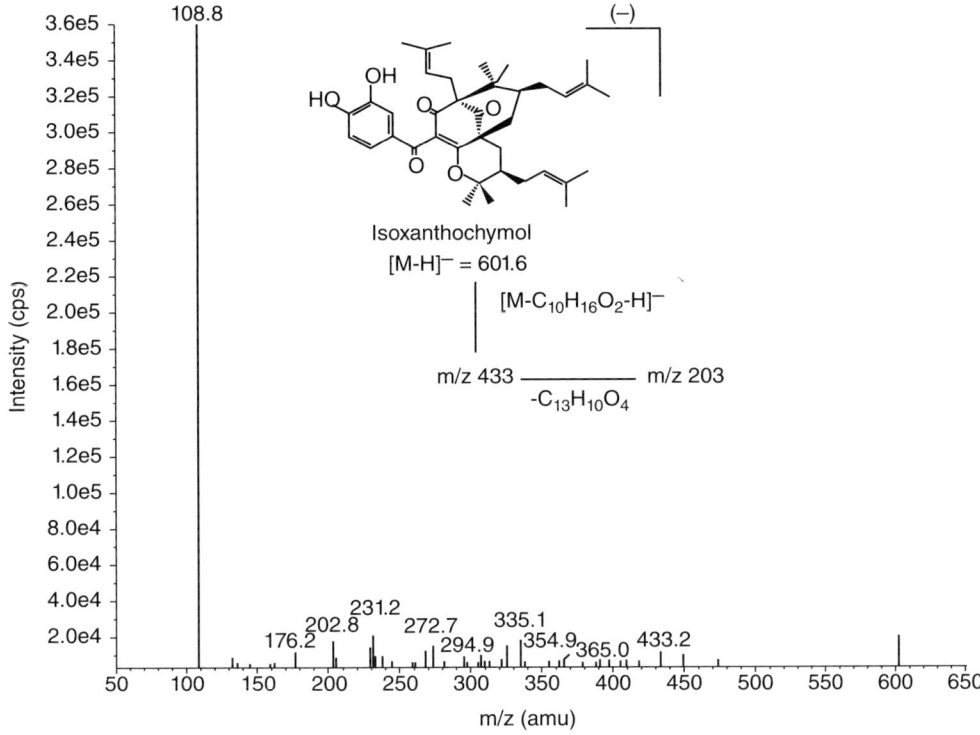

Fig. 6.2b. Product ion mass spectrum (Q2 MS) of isoxanthochymol ($C_{38}H_{50}O_6$) in negative ionization mode. (From Chattopadhyay and Kumar, 2006.)

offer resolution that is only two or three times lower than that provided by high-resolution spectrometers. Hybrid quadruple/time-of-flight mass detectors (Q/TOF-MS) offer high selectivity, high sensitivity and high mass accuracy. The selectivity and sensitivity of TOF-MS detectors make it a good alternative to PDA for determining low UV-sensitive compounds that are difficult to analyse in small amounts (Oszmianski *et al.*, 2013).

Quantitative analysis with LC-MS is still less robust than with other techniques, as the response depends on the ionization process and the method suffers from ionization suppression problems when complex matrices are analysed. It is sensitive and can be considered specific. But the development of bench-top instruments that are easy to operate will continue to increase its use in many areas of natural product research. The structural information obtained from LC-MS is very limited in comparison to that obtained from NMR spectroscopy.

6.5 Liquid Chromatography–Nuclear Magnetic Resonance Spectrometry (LC-NMR)

The recent introduction of HPLC coupled with nuclear magnetic resonance (LC-NMR) represents a powerful complement to LC-UV-MS screening.

HPLC-NMR provides a powerful tool that can yield important complementary information, or in some cases, a complete structural assignment of natural products online (Wolfender *et al.*, 2005, 2006). Ideally, HPLC-NMR should enable the complete structural characterization of any plant metabolite directly in extract, provided that its corresponding liquid chromatographic peak is clearly resolved. However, this is not entirely true as many factors hinder online structure determination. These problems are linked mainly to the inherent low sensitivity of NMR in detecting microgram or submicrogram quantities of natural products separated by conventional HPLC and the need for solvent suppression. Consequently, the limits of detection are several orders of magnitude higher than those of HPLC-UV or HPLC-MS. These limits also depend on the type of magnet used and the type of flow probe employed. The mode of flow (on flow or stop flow) also affects sensitivity strongly. The technique is extremely useful for detecting compounds that are labile or might epimerize or interconvert as a result of their isolation (Cogne *et al.*, 2005; Wolfender, 2008). HPLC-NMR, in the flow mode, can provide ^{1}HNMR spectra of the main constituents of crude extract rapidly without the need for complex automation. This type of information, in addition to other online spectroscopic data (PDA and MS), can be used to ascertain the structural confirmation, and would also be useful in the dereplication process. On-flow HPLC-NMR analysis of flavonoids, alkaloids, terpenes and carotenoids was reported (Jaroszewski, 2005a,b; Wolfender *et al.*, 2006, 2008). Solvent suppression requirement is one of the limiting factors of online HPLC-NMR, as HPLC separation is carried out with acetonitrile–water (MeCN-D$_2$O) and methanol–water (MeOH-D$_2$O). The very large signal of acetonitrile (MeCN) or methanol (MeOH) and the residual water (HOD) signal are completely eliminated by powerful solvent suppression sequences. However, the analyte signal localized under solvent resonance will also be suppressed. Also, the chemical shift recorded in solvent used for reverse-phase HPLC solvent will differ from those reported in standard deuterated NMR solvents. To circumvent the above problems, new approaches allow the use of HPLC-NMR at line in place of online. In this condition, trapping of HPLC signals on solid-phase extraction (SPE) prior to NMR detection can be performed (Exarchou *et al.*, 2005; Jaroszewski, 2005a,b). This includes HPLC microfractionation of the extract, drying and reinjection of the concentrated HPLC in deuterated solvent and using microflow capillary HPLC-NMR probes (Olson *et al.*, 2004; Hu *et al.*, 2005; Glauser *et al.*, 2008). These at-line HPLC-NMR analyses provide a good HPLC peak pre-concentration and high-quality one-dimensional and two-dimensional NMR spectra in fully deuterated solvents without the need for solvent suppression. As compared to the on-flow approach, more sample handling or automation is required in work at line. However, these techniques are crucial for the complete *de novo* structure elucidation of given HPLC peaks, as they can provide ^{1}H-^{13}C correlation experiments in the low microgram range (Wolfender *et al.*, 2005).

SFE-NMR offers the advantage of directly monitoring the extraction process. LC-UV-NMR has great potential for the rapid screening and identification of plant constituents. Albert (1995) reported direct online coupling between separation and extraction techniques such as HPLC, SFC, SFE and CE and proton high-field NMR spectroscopy. SFE-NMR offers the advantage of directly monitoring the extraction process.

The resolution of ^{1}H NMR spectra obtained in HPLC-NMR, SFC-NMR and SFE-NMR coupling under continuous flow conditions was similar to conventionally recorded NMR spectra. The detection limit of an HPLC-NMR separation in acetonitrile-D$_2$O for aliphatic signals of a low molecular mass compound (300 Da) was 500 ng of the injected compound in the continuous flow mode in a 600-MHz NMR spectrometer with a 120-µl flow cell. A similar detection limit in the nanogram range with a 5-nl flow cell was reached in a CE-NMR separation in a 300-MHz NMR spectrometer. However, in CE-NMR coupling, the signal line width degraded but the resolution of the CE-NMR spectra was improving continuously. In SFC-NMR coupling, the whole proton spectral range could be observed without any solvent windows.

In the area of natural products, LC-MS-NMR has been applied as a rapid screening method of searching for unknown marine natural products in chromatographic fractions (Sandvoss *et al.*, 2001) and for the separation and characterization of natural products of plant origin (Bailey *et al.*, 2000; Fritsche *et al.*, 2002). Another application is in the area of combinatorial chemistry (Holt *et al.*, 1997). LC-MS-NMR has been used extensively in the fields of drug metabolism, for the identification of metabolites (Burton *et al.*, 1997; Dear *et al.*, 1998, 2000) and in pharmaceutical research (Pullen *et al.*, 1995).

6.6 Liquid Chromatography–Infrared Spectrometry (LC-IR)

The coupling of a liquid chromatography system with infrared spectrometry as the detection method creates a liquid chromatography–infrared spectrometry (LC-IR) hyphenated technique. The hyphenation is commonly known as LC-IR/Fourier transform infrared spectroscopy (FTIR) or HPLC-IR/FTIR. Infrared spectroscopy (IR/FTIR) provides useful information about organic compounds, as absorption bands in the mid-infrared region are characteristic of functional groups. The combination of HPLC-IR is tedious, and also the sensitivity of IR detection is low as compared to other methods of detection, such as UV–visible and mass spectrometric detection. Two basic approaches, such as the flow cell approach and solvent eliminations, have been attempted to overcome the above limitations. The use of deuterated solvent (D$_2$O/CD$_3$OD) can make it feasible to collect information about C–H bonding. The solvent elimination approach is preferred: here, solvent is evaporated before detection in an infrared detector. Potassium chloride or bromide salts are used for the collection of analytes from the eluents before heating. Diffuse reflectance infrared Fourier transform (DRIFT) and buffer memory techniques are used for solvent elimination approaches (Jinno *et al.*, 1982; Jinno, 2001).

6.7   Capillary Electrophoresis–Mass Spectrometry (CE-MS)

Capillary electrophoresis (CE) was developed from combining several features of different methods, including the principle of gel electrophoresis, the fused silica capillary of gas chromatography and the highly sensitive detectors of HPLC (Suntornsuk, 2007). CE was introduced in the early 1980s. It is an automated analytical and separation technique driven under the influence of the electric field that occurs in a capillary filled with buffer. Separation is based on the differential movement of ions under the applied electric field. The size, charge, degree of ionization, viscosity, etc., of the analyte controls the speed of movement. Three mechanisms are accounted for in separation: (i) differences in analyte mobilities; (ii) differences in analyte isoelectric points; and (iii) differences in analyte conductivity. Instrumentation for CE consists of electrodes, sample introduction systems, a capillary, voltage supply, detector and liquid-handling system. Detection in CE can be performed directly on a capillary. UV-diode array, electrochemical and spectroflurometer are the common online detectors, while a mass spectrometer can be used as an external detector. Liquid-handling systems include an auto sampler, buffer replenishment, buffer levelling and fraction collector.

CE is promising for the separation and analysis of active ingredients in natural products as it needs only small amounts of standards and can analyse samples rapidly. It is also a good tool for generating chemical fingerprints because it has similar technical characteristics to liquid chromatography. Quantitative analysis is done mainly by the comparison of the mobility or migration time of the compounds of interest with those of the standards. Quantitative analysis is calculated from the peak height or peak area based on a calibration curve of a standard. CE is also recommended in several pharmacopeia.

The situation of CE analysis in hyphenation development is somewhat like HPLC analysis. The hyphenated CE instruments such as CE-PDA, CE-MS and CE-NMR have already been reported. Online coupling of CE with MS and other spectrometry allows both efficient separation by CE and specific and sensitive detection to be achieved.

The mass spectra of the separated analyte are recorded using mass spectrometric detection. Selectivity and sensitivity of this hyphenated technique are high. The ESI-MS detection method is generally used in most of the CE-MS techniques, since ESI is considered to be appropriate for providing information about molecular weight and structural characteristics (Dunayevskiy *et al.*, 1996). The coupling of CE with MS creates some complexity as MS cannot accept CE solvents or electrolytes, and conventional gas-phase ionization in MS is not suitable for compounds separated by CE, which are thermally labile, polar or have a high molecular weight. The coupling of CE-MS was introduced successfully in 1988 (Smith and Udseth, 1988). Much progress has been made since then to make CE-MS applicable for analysis of a wide variety of analytes. ESI can be performed by either liquid support or non-liquid support. Sheath liquid is chosen to overcome the volatility and conductivity problems of CE buffer, which usually has poor electrospray capability. Sheath liquid is usually composed of organic solvents (acetonitrile, methanol and 2-propanol) and acidic or basic buffer to promote the ionization of positive and negative ions, respectively. CE can be coupled to different mass analysers such as quadruple, ion trap (IT) and TOF. IT and TOF offer high speed and sensitivity. IT can be used for MS^n experiments, whereas TOF gives high mass resolution and the possibility of measured mass accuracy.

CE is important for drug discovery, manufacturing and other aspects such as determination of active ingredients, enantiomeric separation, etc. Although HPLC is still predominant, CE will continue to serve as a complementary technique in natural product research. It is an ideal technique to be miniaturized on a microchip (Suntornsuk, 2007).

6.8 Conclusion

Smaller, faster and high-resolving instruments for high-throughput analysis in a short time and with low cost are desirable. Hyphenated methods create unprecedented opportunities and their main advantages include extremely low limits of detection and quantification, high precision and repeatability of the determinations. On the other hand, hyphenated methods have their own limitations, such as high price and complexity of the apparatus. Consequently, they are not in general laboratory usage. The application of a hyphenated method requires perfect understanding of analytical methodologies and apparatus. These systems are expensive and are used in scientific research rather than for routine analysis.

References

Albert, K. (1995) On line use of NMR detection in separation chemistry. *Journal of Chromatography A* 703, 123–147.

Bailey, N.J.C., Cooper, P., Hadfield, S.T., Lenz, E.M., Lindon, J.C., *et al.* (2000) Application of directly coupled HPLC-NMR-MS/MS to the identification of metabolites of 5-trifluoromethylpyridone (2-hydroxy-5-trifluoromethylpyridine) in hydroponically grown plants. *Journal of Agricultural and Food Chemistry* 48, 42–46.

Burton, K.I., Everett, J.R., Newman, M.J., Pullen, F.S., Richards, D.S., *et al.* (1997) Determination of SR 49059 in human plasma and urine by LC-APCI/MS/MS. *Journal of Pharmaceutical and Biomedical Analysis* 15, 1903–1912.

Chattopadhyay, S.K. and Kumar, S. (2006) Identification and quantification of two biologically active polyisoprenylated benzophenones xanthochymol and isoxanthochymol in *Garcinia* species using liquid chromatography-tandem mass spectrometry. *Journal of Chromatography B* 844, 67–83.

Chen, Y., Li, Z., Xue, D. and Qi, L. (1987) Determination of volatile constituents of Chinese medicinal herbs by direct vaporization capillary gas chromatography mass spectrometry. *Analytical Chemistry* 59, 744–748.

Cogne, A.L., Queiroz, E.F., Marston, A., Wolfender, J.L., Mavi, S., *et al.* (2005) On-line identification of unstable iridoids from *Jamesbrittenia fodina* by HPLC-MS and HPLC-NMR. *Phytochemical Analysis* 16, 429–439.

Dear, G.J., Ayorton, J., Plumb, R.S, Sweatman, B.C., Ismail, I.M., *et al.* (1998) A rapid and efficient approach to metabolite identification using nuclear magnetic resonance spectroscopy, liquid chromatography/mass spectrometry and liquid chromatography/nuclear magnetic resonance spectroscopy/sequential mass spectrometry. *Rapid Communications in Mass Spectrometry* 12, 2023–2030.

Dear, G.J., Plumb, R.S., Sweatman, B.C., Parry, P.S., Roberts, A.D., *et al.* (2000) Use of directly coupled ion exchange liquid chromatography-nuclear magnetic resonance spectroscopy as a strategy for polar metabolite identification. *Journal of Chromatography B* 748, 295–309.

Delazar, A., Reid, R.G. and Sarkar, S.D. (2004) GC-MS analysis of essential oil of the oleoresins from *Pistacia atlantica* var mutica. *Chemistry of Natural Compounds* 40, 24–27.

Dunayevskiy, Y.M., Vouros, P., Winter, E.A., Shipps, G.W. and Carell, T. (1996) Application of capillary electrophoresis-electrospray ionization spectrometry in the determination of molecular diversity. *Proceedings of the National Academy of Sciences* 93, 6152–6157.

Ellis, L.A. and Roberts, D.J. (1997) Chromatographic and hyphenated methods for elemental speciation analysis in environmental media. *Journal of Chromatography A* 774, 3–19.

Exarchou, V., Krucker, M., Vervoot, J., van Beek, T.A., Gerothanassis, I.P., *et al.* (2005) LC-NMR coupling technology: recent advancements and applications in natural product analysis. *Magnetic Resonance Chemistry* 43, 681–687.

Exarchou, V., Fiamegos, Y.C., van Beek, T.A., Nanos, C. and Vervoot, J. (2006) Hyphenated chromatographic techniques for the rapid screening and identification of antioxidants in methanolic extracts of pharmaceutically used plants. *Journal of Chromatography A* 1112, 293–302.

Fritsche, J., Angoelal, P. and Dachtler, M. (2002) On-line liquid-chromatography-nuclear magnetic resonance spectroscopy-mass spectrometry coupling for the separation and characterization of secoisolariciresinol diglucoside isomers in flaxseed. *Journal of Chromatography A* 972, 195–203.

Ge, G.B., Zhang, Y.Y., Hao, D.C., Hu, Y., Luan, H.W., *et al.* (2008) Chemotaxonomic study of medicinal *Taxus* species with fingerprint and multivariate analysis. *Planta Medica* 74, 773–779.

Glauser, G., Guillarme, D., Grata, E., Boccard, J., Thiocone, A., *et al.* (2008) Optimized liquid-chromatography-mass spectrometry approach for the isolation of minor stress biomarker in plant extracts and their identification by capillary nuclear magnetic resonance. *Journal of Chromatography A* 1180, 90–98.

Herderich, M., Richling, E., Roscher, R., Schneider, C., Schwab, W., *et al.* (1997) Application of atmospheric pressure ionization HPLC-MS-MS for the analysis of natural products. *Chromatographia* 45, 127–132.

Holt, R.M., Newman, M.J., Pullan, F.S., Richards, D.S. and Swanson, A.G. (1997) High-performance liquid chromatography/NMR spectrometry/mass spectrometry: further advances in hyphenated technology. *Journal of Mass Spectrometry* 32, 64–70.

Hu, J.F., Garo, E., Yoo, H.D., Cremin, P.A., Zeng, L., *et al.* (2005) Application of capillary scale NMR for the structure determination of phytochemicals. *Phytochemical Analysis* 16, 127–133.

Jaroszewski, J.W. (2005a) Hyphenated NMR methods in natural products research, part 1: direct hyphenation. *Planta Medica* 71, 691–700.

Jaroszewski, J.W. (2005b) Hyphenated NMR methods in natural products research, part 2: HPLC-SPE-NMR and other new trends in NMR hyphenation. *Planta Medica* 71, 795–802.

Jinno, K. (2001) Infrared detect. In: Cazes, J. (ed.) *Encyclopedia of Chromatography*. Marcel Dekker, New York.

Jinno, K., Fujimoto, C. and Hirata, Y. (1982) An interface for the combination of micro high performance liquid chromatography and infrared spectrometry. *Applied Spectroscopy* 36, 67–69.

Kingston, D.G.I. (1979) High performance liquid-chromatography of natural products. *Journal of Natural Products* 42, 237–260.

Larsen, T.O. and Hansen, M.A.E. (2008) Dereplication and discovery of natural products by UV spectroscopy. In: Colegate, S.M. and Molyneux, R.J. (eds) *Bioactive Natural Products: Detection, Isolation and Structural Determination*, 2nd edn. CRC Press, London, pp. 221–244.

Liang, Z., Jiang, Z., Ho, H. and Zhao, Z. (2007) Comparative analysis of *Oldenlandia diffusa* and its substitutes by high performance liquid chromatography fingerprint and mass spectrometric analysis. *Planta Medica* 73, 1502–1508.

Michalski, R., Jablonska-Czapla, M., Lyko, A. and Szopa, S. (2013) Hyphenated methods for speciation analysis. *Encyclopedia of Analytical Chemistry*, 1–17. DOI: 10.1002/9780470027318. a9291.

Natishan, T.K. (2004) Recent developments of achiral HPLC methods in pharmaceuticals using various detection modes. *Journal of Liquid Chromatography & Related Technologies* 27, 1237–1316.

Niemi, R., Taipale, H., Ahlmark, M., Vepsalaine, N.J. and Jarvinen, T. (1997). Simultaneous determination of clodronate and its partial ester derivatives by ion-pair reversed-phase high-performance liquid chromatography coupled with evaporative light-scattering detection. *Journal of Chromatography B* 701, 97–102.

Olson, D.L., Norcross, J.A., 'Neil-Johnson, M.O., Molitor, P.F., Detlefsen, D.J., Wilson, A.G. and Peck, T.L. (2004) Micro flow NMR: concepts and capabilities. *Analytical Chemistry* 76, 2966–2974.

Oszmianski, J., Kolniak-Ostek, J. and Wojdylo, A. (2013) Application of ultra performance liquid chromatography-photodiode detector-quadrupole/time of flight mass spectrometry (UPLC-PDA-Q/TOF-MS) method for characterization of phenolic compounds of *Lepidium sativum* L. sprouts. *European Food Research Technology* 236, 699–706.

Pullen, F.S., Swanson, A.G., Newman, M.J. and Richards, D.S. (1995) On-line liquid chromatography/nuclear magnetic resonance mass spectrometry – a powerful spectroscopic tool for the analysis of mixtures of pharmaceutical interest. *Rapid Communications in Mass Spectrometry* 9, 1003–1006.

Purcaro, G., Moret, S. and Conte, L. (2012) Hyphenated liquid chromatography-gas chromatography technique: recent evolution and applications. *Journal of Chromatography A* 1255, 100–111.

Sandvoss, M., Weltring, A., Preiss, A., Levsen, K. and Wuensch, G. (2001) Combination of matrix solid-phase dispersion extraction and direct on-line liquid chromatography-nuclear magnetic resonance spectroscopy-tandem mass spectrometry as a new efficient approach for the rapid determination of natural products: application to the total asterosaponin fraction of the starfish *Asterias rubens*. *Journal of Chromatography A* 917, 75–86.

Sarkar, S.D. and Nahar, L. (2012) Natural products isolation. *Methods in Molecular Biology* 864, 1–25.

Shalliker, R.A. (2011) *Hyphenated and Alternative Methods of Detection in Chromatography*. Chromatographic Science Series, Vol. 104. CRC Press, New York.

Smith, R.D. and Udseth, U.D. (1988) Capillary zone electrophoresis-MS. *Nature* 331, 639–640.

Suntornsuk, L. (2007) Capillary electrophoresis in pharmaceutical analysis: a survey on recent applications. *Journal of Chromatographic Sciences* 45, 559–577.

Tswett, M.S. (1906) Physikalisch-Chemische Studien Uber das Chlorophyll. *Die Adsorptionen, Ber. Dtsch Bot. Ges.* 324, 316.

Wilson, I.D. and Brinkman, U.A.T. (2003) Hyphenation and hypernation: the practice and prospects of multiple hypernation. *Journal of Chromatography A* 1000, 325–326.

Wolfender, J.L. (2008) HPLC in natural product analysis: the detection issue. *Planta Medica* 75, 719–734.

Wolfender, J.L., Terreaux, C. and Hostettmann, K. (2000) The importance of LC-MS and LC-NMR in the discovery of new lead compounds from plants. *Pharmaceutical Biology* 38, 41–54.

Wolfender, J.L., Queiroz, E.F. and Hostettmann, K. (2005) Phytochemistry in the microgram domain – a LC-NMR perspective. *Magenetic Resonance Chemistry* 43, 697–709.

Wolfender, J.L., Queiroz, E.F. and Hostettmann, K. (2006) The importance of hyphenated techniques in the discovery of new lead compounds from nature. *Experimental Opinion on Drug Discovery* 1, 237–260.

Wolfender, J.L., Queiroz, E.F. and Hostettmann, K. (2008) Development and application of LC/NMR techniques to the identification of bioactive natural products. In: Colegate, S.M. and Molyneux, R.J. (eds) *Bioactive Natural Products; Detection, Isolation and Structural Determination*. CRC Press, London, pp. 143–190.

7 Non-destructive Techniques

7.1 Introduction

Rapid analytical techniques have been developed as alternatives to the traditional methods of analysis. The application of quick, solvent-free and on-field analytical techniques present an enormous advantage over the conventional wet chemistry approaches (Pedro and Ferreira, 2005). In particular, fast, reliable and non-destructive analysis with minimal sample preparation is required for the quantification of valuable components in screening a large number of samples. These techniques are also finding application in online, semi-continuous monitoring of composition, as sample preparation is minimal in these non-destructive techniques. Furthermore, conventional chemical analysis such as high-performance liquid chromatography (HPLC) or gas chromatography (GC) may take several hours to generate results, which may delay the validation of quality control. Therefore, analytical techniques generating rapid analytical results as compared to conventional methods are required.

7.2 Near-infrared Reflectance Spectroscopy

Near-infrared reflectance (NIR) spectroscopy was first used in agricultural applications by Norris (1964) to measure the moisture in grains. Since then, it has been used for the rapid analysis of many quality traits of a wide variety of agricultural and food products (Davies and Grant, 1987). Recent developments such as multi- and hyperspectral imaging techniques, which provide spatial information, and time-resolved spectroscopy, which allows the measurement of absorption and scattering processes separately, extends the potential of NIR spectroscopy further (Martinsen and Schaare, 1998; Cubeddu *et al.*, 2001; Lu, 2003).

NIR spectroscopy (NIRS) is a multi-trait analysis technique with large application in the analysis of quality characteristics in food and agricultural commodities. NIRS is a valuable tool for monitoring the quality of oilseeds using a few intact seed samples that can be further used for breeding purposes in a quality improvement programme. Use of Fourier transform near-infrared spectroscopy (FT-NIR/NIR) as an analytical technique for oilseed quality monitoring in research laboratories, as well in processing industries, will promote the avoidance of many hazardous chemicals, thereby providing 'a green analytical technique'.

NIRS offers a practical solution for most common quality analysis/control requirements. It provides quick and reliable identification of raw materials as they are received and can be used for accurate quantitative analysis of multi-component mixtures. It provides fast measurement without the need for chemicals and sample preparation and is therefore being used for a wide range of quality control/quality analysis applications in many industries. The development of NIRS revolutionized evaluation of the quality traits of many agricultural commodities.

Infrared spectroscopy is based on the interaction of electromagnetic radiation with matter. Infrared spectroscopy is considered a valuable tool for the qualitative and quantitative investigation of unknown samples. In NIRS, the product is irradiated with NIR radiation, and when the NIR radiation hits a sample, the incident radiation may be reflected, absorbed or transmitted. The relative contribution of each phenomenon depends on the chemical constitution and physical parameters of the sample. The spectral characteristics of irradiated radiation change when it penetrates the sample. This change is dependent on the microstructure of the sample. Microstructure is related to the chemical composition. Every solid, liquid or gaseous material has a periodic lattice of molecules or atoms. The local or global dipole moment is generated by the lattice interaction with the applied electromagnetic radiation. The make-up of the molecule, that is, mass, bonding force, distance and relative angle between the atoms, produces a characteristic vibrational frequency. In contact with electromagnetic radiation, energy corresponding to the vibrational frequency is absorbed by the group or molecule. This produces attenuation of the electromagnetic radiation at these frequencies. The infrared spectrum, a comparison of the incoming electromagnetic radiation with the attenuated signal, is characteristic of the investigated substance. Absorption strength depends directly on the absorption coefficient and the amount IR radiation interacts. A component with 0.1% or more concentration could be detected and evaluated by NIRS. To find a correlation between analyte concentration and IR absorption, a series of samples containing a known amount of analytes are scanned. An equation can be developed with the mathematical correlation to determine the concentration of the analyte. Spectral information can be recorded in both transmittance and absorbance mode. However, transmittance mode requires a more powerful light source and a sensitive detector (Lu and Ariana, 2002). Further, the transmittance mode is limited to 600–1000 nm (0.6–1.0 µm), and most biological materials have stronger absorption beyond this spectral region (Clark et al., 2003).

Lu and Ariana (2002) reported that the transmittance mode might provide more spectral information about the interior properties of fruit, as light travels further in fruit. This mode is suitable for fruit with high transmittance, such as citrus, tomato and small seeds (Greensill and Walsh, 2000). Transmission techniques can be applied for edible oilseeds because many difficulties associated with reflectance spectroscopy, especially related to calibration development and maintenance, are eliminated.

The NIR region meets the visible at about 12,500 cm^{-1} (0.80 µm) and extends to about 4000 cm^{-1} (2.50 µm), thereby covering the wavelength range adjacent to the mid-infrared and extending up to the visible region. The American Society of Testing and Materials (ASTM) defines the NIR region of the electromagnetic spectrum as the wavelength range of 780–2526 nm, corresponding to the wave number range 12,820–3959 cm^{-1}. The NIR technique is based on the correlation between chemical properties, as determined by standard methods and the absorption of light. The NIR spectrum depends on the chemical composition of the samples and the physical characteristics of the samples (Chen *et al.*, 2006). Although the absorptivity of NIR bands is about 10–1000 times lower than that of mid-infrared bands, thicker sample layers (0.5–10 mm) compensate for these smaller molar absorptivities of NIR. Because of the lower absorptivities, the NIR beam penetrates deeper into the samples as compared to ultraviolet, visual or infrared radiation, giving a more representative analysis. However, the low absorption coefficient permits good penetration depth and an adjustment of sample thickness. This aspect is actually an analytical advantage, as it allows direct analysis of strongly absorbing and even highly scattering samples like turbid liquids or solids, either in transmittance or reflectance mode without pre-treatment. Furthermore, minor impurities are less troublesome in both reflectance and transmission methods than in the mid-infrared region. As an analytical tool, mid FT-IR transmission spectroscopy is also gaining acceptance in the edible oilseed sector; however, NIR instruments are more robust, less energy limited and have more sensitive detectors than mid-FTIR spectrometers. Also, glass and quartz cells can be used for recording spectra. Thus, NIR analysis is more suitable than mid-IR analysis, which employs transmission IR cells with narrow path lengths (typically 0.025 mm) and salt windows. NIR analysis can also be performed remotely using NIR transmitting fibre optics (Li *et al.*, 2000). For these reasons, NIR spectroscopy has been used more in industrial applications, in both at-line and online applications.

Most absorption bands in the NIR region are overtone or combination bands of the fundamental absorption bands in the infrared region of the electromagnetic spectrum, which are due to the vibrational and rotational transitions and encompass 'bond vibration' and combination of overtones of the fundamental C-H, O-H and N-H bonds. Combination bands in 1900–2500 nm are due to vibrational interactions, i.e. frequencies are the sums of multiples of each interacting frequency. These bonds are the primary structural components of organic molecules in food and oilseed crops. Different oilseeds have different wavelengths assigned to oil bands due to the interactions between constituents, particle size, chain length and degree of saturation or unsaturation of fatty acids (Bhatty, 1991). Basically, fats and oils are made up of fatty acids esterified with trihydroxy alcohol (glycerol) with a varying number of carbon chains, positions of varying double bonds and fatty acids. Peaks in the regions of 3007 cm^{-1} are attributed to *cis*-C=H vibration and of 1654 cm^{-1} are due to *cis*-C=C vibration. The more unsaturated the fatty acid, the higher the peak intensities at 3007 and 1654 cm^{-1} (Guillen and Cabo, 1997). Many absorption bands are often associated with hydrogen atoms, like the first overtones of the O-H and N-H stretching vibrations near 7140 cm^{-1} (1.40 µm) and 6667 cm^{-1} (1.50 µm), respectively, and combination bands resulting from C-H stretching and deformation vibrations of alkyl groups at 4548 cm^{-1} (2.20 µm) and 3850 cm^{-1} (2.60 µm) are also important. Overtones and combination bands are weaker (10–100 times) than the fundamental absorptions occurring in the mid-infrared region. Because of this, in mid-infrared spectra, signals of typical bonds can be observed visually, while in the infrared region, signals are usually not associated with typical bonds.

NIR spectra have a broad envelope with few sharp peaks formed as a result of multiple bands and peak broadening. Spectral profiles in the NIR region contain less spectral detail as it consists of overlapping and poorly defined overtones and combination bands of fundamental bands. However, because of these bands, functional groups are well differentiated in the NIR region, making this region more suitable for quantitative analysis.

7.2.1   Instrumentation

An NIR spectrometer consists of a light source, sample presentation accessory, monochromator, detector and optical components such as lenses, collimators, beam splitters, integrating spheres and optical fibres. Three types of NIR spectrometer are presently in use: (i) dispersive; (ii) Fourier transform; and (iii) diode array. Light source (the tungsten halogen lamp is the most common), dispersive unit (filters/gratings/interferometer) and detector (silicon/lead sulfide/indium-gallium arsenide) are the most important parts, and NIR spectrometers are classified according to the type of monochromator. In a filter instrument, the monochromator is a wheel holding a number of absorption or interference filters. Its spectral resolution is limited. A grating or prism is used for separating the individual entering or leaving frequencies in scanning monochromators. The wavelength separator rotates so that the radiation of the individual wavelengths subsequently reaches the detector.

An interferometer is used in Fourier transform spectrometers to generate modulated light: the time domain signal of the light reflected or transmitted by the sample can be converted into a spectrum via a fast Fourier transform. Often, a Michelson interferometer is used, but also polarization interferometers are used in some spectrometers. In photodiode array (PDA) spectrometers, a fixed grating focuses the dispersed radiation on to an array of silicon (350–1100 nm) or InGaAs (indium gallium arsenide, 1100–2500 nm) photodiode detectors. Laser-based instruments have different laser light sources, or a tunable laser, but do not contain a monochromator. Acoustic optic tunable filter (AOTF) instruments use a diffraction-based optical band pass filter that can be tuned rapidly to pass various wavelenths of light by varying the frequency of an acoustic wave propagating through an anisotropic crystal medium. Liquid crystal tunable filter instruments use a birefringent filter to create constructive and destructive interference based on the retardation in the phase between the ordinary and extraordinary light rays passing through a liquid crystal. In this way, they act as an interference filter to pass a single wavelength of light. By combining several electronically tunable stages in series, high spectral resolution can be achieved (Stratis *et al.*, 2001; Nicolai *et al.*, 2007). There is a shift towards PDA systems because of their high acquisition speed (the integration time is typically 50 ms, but can be as low as a few milliseconds).

Dispersive NIR spectrometers can be unreliable in terms of maintaining calibration stability (Li *et al.*, 2000). FT-NIR spectrometers based on interferometery are an advancement over dispersion-based instruments. This technique provides enhanced energy throughput, better signal-to-noise ratio, excellent wave number reproducibility, extensive data manipulative capabilities, accuracy and advance chemometric capabilities for calibration development. Fourier transform–NIR spectrometers are based on the mathematical operations known as Fourier transformations (FT), a powerful method to enhance the signal and noise ratio. With the combination of FT and chemometrics, the whole spectrum is recorded instantaneously and simultaneously. Here, data are collected rapidly in the time domain and then converted by FT to the conventional frequency domain. As the time required for a single scan is reduced, the time required for ensemble averaging is also reduced. FT-NIR requires a small quantity of sample in comparison to conventional dispersive NIR spectrometers.

7.2.2 Calibration development

The selection of pre-analysed samples, recording their spectra and the development and validation of calibration models are three important steps in method development using the NIR/FT-NIR technique. Both intact and ground samples are used in NIR spectrometers. The important impact of grinding is to obtain a fine and homogeneous powder. However, a long period of grinding leads to the destruction of oil bodies, as well as an increase in temperature of the sample, forming conglomerates that cannot be eliminated by mixing. Therefore, for reproducible results, the use of a uniform grinding protocol is necessary. Based on the optical response of the samples – that is, the absorption and scattering characteristics – NIR measurements are done either in transmittance or in reflectance mode, depending on their absorption and scattering characteristics. Different modes of measurement are shown in Fig. 7.1. Transparent materials are

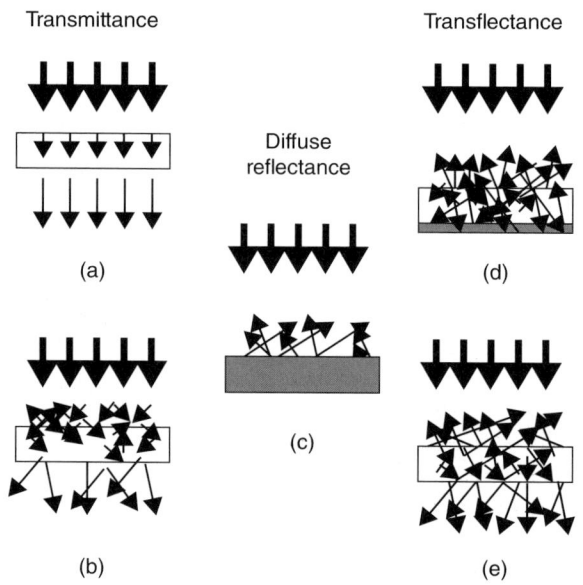

Fig. 7.1. Different modes of measurement used in NIR: (a–b) transmittance; (c) diffuse reflectance; and (d–e) transflectance.

usually measured in transmittance mode (Fig. 7.1a). Turbid liquids or semi-solids and solids may be measured either in diffuse transmittance (Fig. 7.1b), diffuse reflectance (Fig. 7.1c) or transflectance (Fig. 7.1d–e). However, in all models, absorbance (A) values relative to a standard reference material are measured with A corresponding to log 1/R and log 1/T for reflectance and transmittance spectra, respectively (Reich, 2005).

For recording spectra, the samples are scanned in a sample cup, either in transmittance mode or reflectance mode, in the region of 800–2500 nm at 1- or 2-nm intervals (Figs 7.2 and 7.3).

Each recorded spectra is an average of 16/32/64 or more scans (Fig. 7.4). Each sample is scanned twice and the two spectra are averaged. Recorded spectra are subtracted against the background air spectrum. Reflectance data are stored as the logarithm of reflectance; log 1/R. The best spectral region is selected for the optimization of spectral processing, manipulation and preparation of the calibration curve. Water is the most important chemical constituent of most agricultural produce. As water highly absorbs NIR radiation, the NIR spectrum of agricultural produce is dominated by water. Further, the NIR spectrum is essentially composed of a large set of overtones and combination bands. This, in combination with the complex chemical composition of the agricultural produce, causes the NIR to be highly complicated. Wavelength-dependent scattering effects, tissue heterogeneities, instrumental noise, ambient effects and other sources of variability also complicate the spectrum. Multivariate statistical techniques, also called chemometrics, are therefore required to extract information about the quality attributes buried in the NIR spectrum.

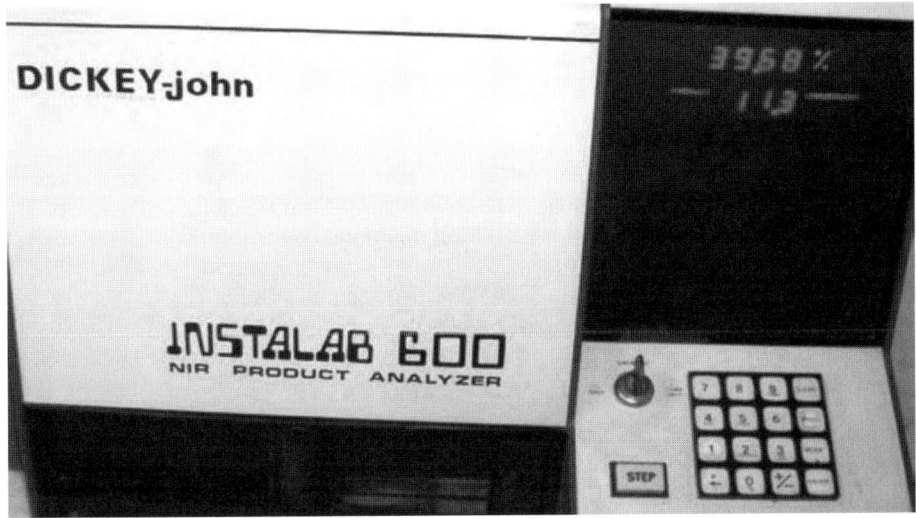

Fig. 7.2. NIR product analyser (Dickey John, USA).

Fig. 7.3. NIR sample cup filled with Indian mustard (*Brassica juncea*) seed.

Regression techniques are essentially used for this, along with spectral preprocessing (Geladi and Dabakk, 1995).

Spectral preprocessing techniques are used to remove any irrevalent information that cannot be handled properly by regression techniques. Original reflectance spectra are corrected prior to calibration by applying mathematical transformations like scattering correction, vector normalization, multiplicative scattering correction, min-max normalization, constant offset elimination, first derivative and second derivative, etc., either alone or in combination, to find out if they would improve the performance of

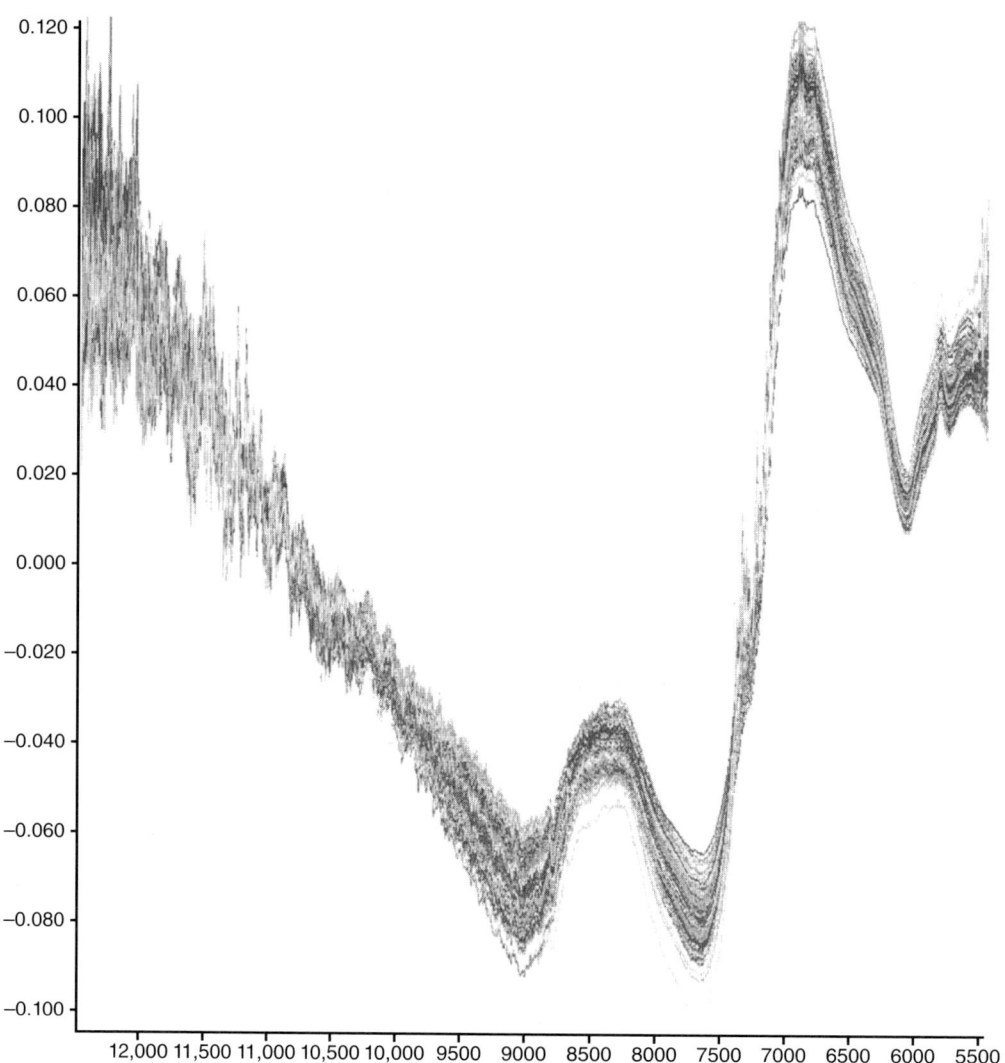

Fig. 7.4. FT-NIR spectra of intact safflower seed recorded in reflectance mode using Matrix-I FT-NIR spectrometer from Brucker Optics, Germany.

the calibration model. Absorption bands in the NIR region overlap very extensively; therefore, traditional univariate analysis techniques are generally not applicable and advanced chemometric techniques are used to obtain meaningful data. Interference and overlapping of the spectral information are overcome by using powerful multi-component analysis such as partial least squares (PLS) (Fuller and Griffiths, 1978; Liang and Kvalheim, 1996). This method allows a statistical approach using the full spectral region rather than unique and isolated bands (Galtier *et al.*, 2007). The algorithm is based on the ability to correlate spectral data mathematically to a property matrix of interest while accounting for all other significant spectral factors that perturb the spectrum.

Spectral information and chemometric techniques are employed for calibration development using various algorithms. Chemometrics generates correlations between the experimental data and the chemical composition or physical properties of the tested samples through mathematical and statistical procedures. The analysis software is designed for the quantitative analysis of spectra consisting of bands showing considerable overlap, thereby enabling the simultaneous determination of more than one component in each sample. The spectral data and the chemical data are written in the form of matrices. Each row in the matrix represents a sample spectrum. Further, the matrices are broken down into several factors, called principal components or PLS vectors (Yang and Ren, 2008). Regression is derived by establishing the relationship between the reference data and a set of samples with known composition. Chemometric techniques such as multivariate analysis are used to derive information from the spectra. Multivariate analysis is required, as interpretation models based on overtone bands are more complex than fundamental bands in terms of interpretation. As multivariate NIR spectral data have a huge number of correlated variables, the reduction of variables, that is, to describe data variability by a few uncorrelated variables containing the relevant information for calibration modelling, is required. Using multivariate regression, a relationship between the $nX1$ vector of observed response value y (Y-variables, that is, quality attributes of interest) and the nXN spectral matrix ('X-variables') with n number of spectra and N number of wavelengths is established. In multiple linear regression (MLR), y is approximated by a linear combination of the selected values at every single wavelength. The N regression coefficients are estimated by minimizing the error between the predicted and observed response values in a least square sense. A stepwise MLR may be applied to select a number of variables for the equation. MLR models typically do not perform well due to the high co-linearity of the spectra leading to overfitting of calibration models (Naes *et al.*, 2004).

PLS and principal component analysis (PCA) are advanced multivariate calibration models. PLS is a sophisticated multivariate analysis technique for discriminate and quantitative analysis. Using PLS, classification is through the setting of discrimination criteria based on the output obtained from calibration models. The key difference between PLS and multiple linear regression is that PLS calibration does not entail establishing a direct relationship between concentration and absorbance measurements at specified frequencies. Rather, it develops a model by compressing the spectral data for a set of calibration standards into a series of mathematical spectra, and treats concentration instead of spectral intensity as an independent variable. In principle, a PLS calibration can be based on the whole spectrum; however, in practice, the analysis is restricted to the regions of the spectrum exhibiting the strongest variations with changes in the concentrations of the components of interest. PLS decomposes the spectrum of each calibration standard into a weighted sum of the loading spectra, and weights given to each loading spectrum, known as scores, are regressed against concentration data for the standards. In the case of unknown samples, PLS reconstructs the spectrum from the loading spectra. The scores, that is, the amount of each loading spectrum employed in reconstructing the spectrum, are used to predict concentration (Li *et al.*, 1999). In this way, the unidentified sources of spectral interferences (e.g. overlapping bands) common in NIR spectroscopy can be compensated for. However, PLS calibration models only give accurate predictions for samples that are well represented by the calibration standards. Also, in the case of multi-component analysis, the predictions obtained from a PLS model reflect any intercorrelation that exists in the calibration set.

PCA, used for discriminant qualitative analysis, is the best known and most widely used variable reduction method (Li *et al.*, 1999; Reich, 2005). PCA reduces the number of variables to a limited number of principal components (PCs). Unlike measurable variables, PCs are orthogonal, and thereby they describe independent variation structures in the data. The first PC always explains the greatest part and the following PCs successively explain the smaller parts of the orthogonal variance. The presence of significant PCs indicates structure in the data. Graphic overviews, ideally showing a large part of the variance in two dimensions of the objects and variables, are obtained by score and loading plots, respectively (Jolliffe, 1986; Wold *et al.*, 1987; Kamal-Eldin and Andersson, 1997). PCs are linear combinations of the orthogonal variables and are determined so that the first PC explains the largest part of the total variance. This means that correlated variables are explained by the same PCs and less correlated by different PCs.

For qualitative evaluation, discriminate partial least squares analysis (DPLS), an application of PLS for discrimination analysis, has been reported (Moschner and Biskupek-Korell, 2006). Principal component regression is a two-step procedure. Using PCA, first X is decomposed. Then, using a small number of PCs or latent variables, it is fitted in a multiple linear regression model (MLR model). The advantage, in the context of MLR, is that the X variables (PCs) are uncorrected and the noise is filtered. However, its demerit is that PCs are ordered according to the decreasing explained variance of the selected matrix, and the first PCs that were used for the regression model are not necessarily the most informative with respect to the response variables. PLS was introduced to overcome this disadvantage (Wold *et al.*, 2001). Here, an orthogonal basis of latent variables is constructed one by one in such a way that they are oriented along the directions of maximal covariance between the spectral X and response vector Y. This ensures that latent variables are ordered according to their relevance in predicting the Y variables. The interpretation of the relationship between the X data and the Y data (the regression model) is then simplified, as this relationship is concentrated on the smallest possible number of latent variables. In the case of large numbers of correlation or even co-linearity, this model performs well. Also, in comparison to the principal component regression (PCR) calibration model, the required number of latent variables is smaller for a similar model performance. PLS regression can be extended easily to predict several qualities of attribute simultaneously (Naes *et al.*, 2004).

Quantitative analysis by NIRS requires sufficient spectra to make sample measurement sets and validation sets. Calibration models are developed using cross-validation by splitting all the samples randomly into two groups, namely calibration samples and validation samples. Cross-validation is useful to check the performance of the developed calibration equations and it also prevents overfitting (Shenk and Westerhaus, 1993). As the ability of the NIRS model to discriminate or identify similar or different parameters is based on the vibrational responses of chemical bonds to NIR radiation, the higher the variability in the constituents, the better the accuracy of the model in responding to electromagnetic NIR radiation. In leave-one-out cross-validation, the first sample is removed from the calibration data set and the remaining samples are used to find the regression model. Subsequently, the omitted sample is predicted using the new regression. The procedure is repeated, leaving each sample out in turn. Then, for each calibration sample, the difference between the true and the predicted value is calculated. The sum of the square of the differences is calculated. The lower the value, the better the predictive capability of the developed model (Miller and Miller, 2005). In multifold cross-validation, a well-defined number of samples (a segment) are left out instead of one. In internal validation, the data set is spilt into a calibration and a validation set. The calibration model is constructed using the calibration set; the prediction residuals are then calculated by applying the calibration model to the validation set. In external validation, the validation data set is independent. For example, samples obtained from different seasons or locations are used. To avoid overfitting of the calibration model, as well as to assess accuracy, validation procedures are applied. Leverage correction is an equation-based procedure to establish the prediction accuracy without performing any prediction. It leads to overoptimistic estimates; therefore, it is avoided at all times.

Different mathematical treatments, like the stepwise elimination of outliers, are also done to improve the accuracy of the calibration curve. Multiple coefficients of determination of calibration (R^2), coefficients of determination of external validation (r^2), standard error of cross-validation (RMSECV), standard error of performance (SEP) and residual prediction deviation (RPD) are important parameters used to judge the quality of the model developed. At the same time, the suitability of a model for screening is also judged using the above-mentioned parameters. The relationship between the measured and predicted vales is expressed by the regression coefficient (r), and it describes the quality of quantitative calibration (Williams, 1987). A value of r greater than 0.91 suggests a good correlation. The value of R^2 shows the proportion of the variance in reference data that can be explained by the variance in the predicted data. When it is greater than 0.83, the robustness of the prediction of the calibration model is maintained (Elfadl *et al.*, 2010). Once developed, the calibration curve is utilized to validate the quality parameter in unknown samples. Ideally, the sample set for calibration and validation should have uniform distribution of composition across the anticipated range. Further, statistical tests like the student *t*-test, etc., are applied to compare the results using FT-NIRS and the reference method.

Data preprocessing is an important stage in performing a calibration, and spectra are improved significantly with optimal preprocessing. In quantitative analysis, it is assumed that the layer thickness is the same in all measurements; however, sample thickness varies in operation. These variations can be eliminated to ensure a good correlation between the spectral data and the chemical values. The number of PLS vectors is a crucial point in the quality of the calibration model. The concentration range to be covered by the samples and the distribution within this range should be thoroughly considered in developing a calibration set of optimum size, as spectral interference may degrade the performance of the calibration model when the sample size is too large (Centner *et al.*, 1996). Also, the cost and time spent in the reference analysis of large samples are not often affordable. However, the sample used should represent the whole spectral and chemical variability in the calibration and validation groups. The optimized calibration model needs to be validated using a validation group of samples analysed for components using a reference method. In order to validate the calibration model, cross-validation is accomplished by a leave-one-out technique.

7.2.3   Spectral preprocessing methods

After collecting sufficient spectra, the two major steps involved are selecting the wavelength range and preprocessing the spectra (Pang *et al.*, 2008). Selecting the limited wavelength range is important because information derived from the full wavelength range is enormous and the absence of signals in certain regions of the spectra may influence the accuracy of the results. Therefore, it is important to select the wavelength range of positive correlation and reject the range of negative correlation.

Preprocessing spectra is a procedure to optimize data and avoid disturbance due to a changing baseline.

Different processing techniques are applied to recorded spectra to find out if they would improve the performance of the calibration models. First, the relative spectrum is used without any preprocessing, followed by spectrum preprocessing techniques applied on the spectra one at a time. Multiplicative scatter correction (MSC) and standard normal variate (SNV) are mathematical treatments used to compensate for scatter-induced baseline offsets. Both methods developed for reflectance spectra can also be applied to transmittance spectra. Normalization algorithms may reduce or eliminate baseline shifts and intensity differences resulting from variable positioning or path length variations. The resolution of overlapping bands can be improved by the use of derivative spectral preprocessing. In addition, it is able to reduce baseline offsets. Since spectral noise is also amplified by converting into derivatives, derivatives are usually combined with Taylor or Savitzky Golay smoothing algorithms (Savitzky and Golay, 1964; Reich, 2005). Spectral preprocessing methods are described briefly below.

Constant offset elimination

This method can minimize the baseline shift and also enhances the intrinsic matrix absorption of the material, thereby enabling the extraction of significant information from the original spectra matrix.

Multiplicative scatter correction (MSC)

Sometimes, a parallel shift of the spectra is observed due to the influence of particle size. To eliminate the scatter, data treatment with multiplicative scatter correction is performed. MSC was used to modify the additive and multiplicative effects in the spectra (He *et al.*, 2007).

First and second derivative spectra

First derivative spectral preprocessing minimizes the shift in baseline and omits the intensity effect encountered in FTIR spectra. Second derivative removes the slope effect. However, the derivative treatment can influence the measurement sensitivities strongly and therefore should be avoided if the concentrations of the analytes of interest are very low (Cadet and de la Guardia, 2001).

Standard normal vector (SNV)

Preprocessing of spectra with standard normal vectors scales spectral data in order to compensate for path length differences (Wang *et al.*, 2006). Scatter correction of the standard normal variate and detrend transformation (SNV-DT) is usually applied for a better correlation (Barnes *et al.*, 1989). SNV is a mathematical transformation method of the log (1/R) spectra used to remove slope variation and to correct for scatter effects (Chen *et al.*, 2006).

Min-max normalization

If the original spectrum shows slight drift, it can be corrected using min-max normalization.

Straight-line subtraction

Baseline offset can be corrected through straight-line subtraction preprocessing.

Finally, processing techniques, namely first derivative and straight-line subtraction, first derivative and vector normalization and first derivative and multiplicative scattering corrections, are applied one after another.

Smoothing

This is done using the Norris derivative smoothing filter (segment length = 5 nm, gap size = 5 points) (Chen *et al.*, 2012).

7.2.4   Validation parameters

Once developed, the calibration is validated in terms of coefficient of determination (R^2), root mean square error of cross validation (RMSECV), standard error of prediction (SEP) and residual prediction deviation (RPD).

Coefficient of determination (R^2)

The coefficient of determination (R^2) shows the proportion of the variance in the reference data that can be explained by the variance in the predicted data. The coefficient of determination, also called the square of correlation coefficient, and bias, is used to express the predictive ability of the developed calibration. The systematic difference between the predicted and the measured values is called bias and is computed as the average value of the residuals. The measure of the variation not taken into account by the model is residual. For a given sample and a given variable, the residual is computed as the difference between the observed value and the fitted (projected or predicted) value of the variable on the sample.

Root mean square error of cross-validation (RMSECV)

The RMSECV is the prediction error of the calibration model when cross-validation is used. It is defined as the standard deviation of differences between the spectral data and the reference values in the cross-validation sample set. It is a quantitative measure of the preciseness with which the samples are predicted during validation. Optimum calibration should be selected by minimizing the RMSECV value (Sinelli *et al.*, 2008).

Standard error of prediction (SEP)

The standard error of prediction (SEP) gives an estimation of the prediction performance during the step of validating the calibration equation. The standard error of estimation (SEE) and SEP represent the precision of calibration, their ratio, consistency should be as close to 100 as possible (Elfadl *et al.*, 2010).

Residual prediction deviation (RPD)

Residual prediction deviation (RPD) is a qualitative measurement of the assessment of validation results. The larger the RPD, the greater the probability of the model to predict accurately. The RPD is defined as the ratio of the standard deviation of the response variables to the RMSEP or RMSCV. An RPD between 1.5 and 2 means that the model can discriminate the low from the high values of the response variables. A value between 2 and 2.5 indicates that coarse quantitative predictions are possible and a value between 2.5 and 3 or above corresponds to good and excellent prediction accuracy, respectively (Williams and Norris, 2001). Lower values can result from the narrow range of reference values (Cozzolino *et al.*, 2008).

Improving the calibration model by correction methods is crucial for accurate quantitative analysis. The stability of the mathematical model dictates the accuracy of quantitative analysis by NIRS. It is influenced by a series of factors such as the accuracy of chemical reference values, the number of spectra representing the model, equipment factors and user's factors. A sufficient number of representative samples, strict control of experimental conditions and adoption of an appropriate data processing method all have an impact on the chemical reference values and accuracy.

In robust calibration models, accuracy is relatively unaffected by unknown changes of external factors. The following factors may affect the performance of the model (Wang *et al.*, 1991):

1. When the calibration model developed on one instrument is transferred to another instrument, the response of each instrument may differ.
2. Because of temperature fluctuations, electronic drift and changes in wavelength or detector stability over time, instrumental response may differ.
3. Samples represent different batches. This factor is the most important factor in the case of fruit and vegetable samples. Their matrix may show variability due to the age of the tree, crop load, spur age, position within the tree and light effects. Variability may also be due to soil characteristics, nutrition and weather conditions, fruit age and seasonal variability (Peirs *et al.*, 2002b).

Appropriate external validation is of prime importance for the successful application of multivariate calibration models. Lack of robustness gets transferred into bias. In order to obtain a sufficiently robust calibration model, the calibration data set should be sufficiently rich in variation and contain samples from different locations and seasons.

Spectrophotometers can vary both in wavelength calibrations and photometric response because of manufacturing tolerances, differences in optics, detectors and light sources, changes in the instrument response function over time (ageing of sources, replacement of some part, etc.) and changes in the instrument's environment over time, such as temperature, humidity, etc. (Greensill and Walsh, 2000; Fearn, 2001). Therefore, calibration models developed for one instrument are, as such, useless on another instrument, even within one model. Preprocessing of the spectral data (4000–1200 cm^{-1}) of safflower seed recorded using FT-NIRS is described in Table 7.1 (Kumar and Andy, 2013).

Table 7.1. Optimization of spectral processing method for estimation of the oil content (%) in intact safflower seeds using FT-NIRS.

Spectral preprocessing method	Rank	RMSECV	RPD	R^2	Spectral range (cm^{-1})
Multiple scattering	5	0.413	0.312	0.889	12,493.3–4,246.7
First derivative	5	0.709	1.75	0.675	7,502.1–5,446.3
Second derivative	7	0.629	1.98	0.744	6,102.0–4,246.7
Straight-line subtraction	6	0.277	4.49	0.950	12,493.3–5,446.3
Constant offset elimination	6	0.280	4.49	0.909	12,493.3–5,446.3
Vector normalization	5	0.413	3.02	0.889	5,450.1–4,597.7
Min-max normalization	5	0.397	3.13	0.898	12,493.3–4,597.7
First derivative + straight-line subtraction	5	0.652	1.91	0.723	7,502.1–5,446.3
First derivative + vector normalization	5	0.710	1.75	0.674	7,502.1–4,597.7
First derivative + multiple scattering	5	0.070	1.77	0.681	7,502.1–5,446.3
No spectral processing	3	2	1.24	0.123	12,493.3–5,446.3

Commercial implementations of NIR spectroscopy require the transferability of the calibration model from one instrument to another. This will provide a reduction in the cost of repeating the calibration process. To transfer a calibration model developed on a master or primary instrument to another instrument, called slave or secondary, a number of calibration transfer techniques, also known as instrument standardization, have been developed. Spectra on the slave instrument are transferred so as to appear as if originating from the master instrument. The original calibration can be used on the transferred spectra without further calibration.

An NIR spectrum incorporates a large amount of information and includes overlapping and interconnected signals. Pattern recognition is an important approach to reduce the number of variables. Pattern recognition is categorized as unsupervised or supervised (Li *et al.*, 2007). Cluster analysis, PCA and discriminant analysis are categorized into the unsupervised class. Latent projection, *k*-nearest neighbour algorithm (KNN), Fisher linear discriminant analysis (FLDA) and artificial neural networks (ANNs) belong to the supervised class.

Cluster analysis

Cluster analysis is a convenient method to simplify data from multi-component systems for data analysis and data mining. In the case of multi-component systems where the samples give distinct differences in the NIR spectra, samples can be classified using cluster analysis (Liu *et al.*, 2007; Guo *et al.*, 2009).

Principal component analysis (PCA)

Principal component analysis (PCA) is a mathematical procedure to reduce the dimension of data by linear fitting (Roggo *et al.*, 2003). It produces a group of new variables that represent the primary information of the original variables without any loss of data (Yuan *et al.*, 2011). The dimensions of the data are reduced linearly to calculate cumulative contribution rates and score plots (Wang *et al.*, 2009). The results are presented on two- or three-dimensional maps to demonstrate the quality.

K-nearest neighbour algorithm (KNN)

K-nearest neighbour algorithm (KNN) is a non-parametric method for classifying objects based on the closest training examples in the feature space. It ranks the contributions of neighbours in terms of their closeness to the object (Li *et al.*, 2012).

Latent projection

Methods based on quantitative calibration and pattern recognition are established on the basis of latent projection. PLS, partial least squares discriminant analysis (PLS-DA) and soft independent modelling of class analogy (SIMCA) were developed from latent projection (Lucio-Gutierrez *et al.*, 2011; Lu *et al.*, 2012; Fu *et al.*, 2013).

Artificial neural networks

Artificial neural networks (ANNs) are a new processing method having self-adaptive and massively parallel machine learning systems composed of layers of processing elements called neurons (Zhang *et al.*, 2007). Here, many simple neurons are connected to each other to make a complex network. ANNs are used primarily for solving pattern recognition problems by building non-linear models. The models can be used to generate their conclusions and to predict patterns that have not been been encountered previously. This complex network is used to calculate values by feeding information through the network (Warren and Pitts, 1943). Radial basis function ANN (RB-ANN) and back propagation ANN (BP-ANN) are common applications used to overcome some of the disadvantages of NIRS, such as broad spectral range, poor signal strength and overlapping signals (Qi *et al.*, 2003). ANNs reduce interference and noise and have good non-linear conversion capabilities to avoid prediction error effectively (Hou *et al.*, 2001). The combination of NIRS and ANNs is particularly useful for identifying plant samples with indistinguishable features (Hou *et al.*, 2001; Qi *et al.*, 2003; Gao *et al.*, 2006). Also, knowledge plasticity to changing inputs and outputs, fault tolerance, noise immunity and interpolation/extrapolation capabilities are the particular advantages of ANNs (Zhang *et al.*, 2007).

Specific spectrum method

Direct comparison of a sample spectrum with the spectrum of an authentic specimen is an important method for qualitative analysis. It is similar to fingerprinting as applied in chromatography and spectroscopy (Zhou *et al.*, 2008; Ni *et al.*, 2009; Tian *et al.*, 2009).

7.2.5   Application of NIRS for the identification and quantification of bioactive compounds from medicinal plants

NIRS is one of the fastest growing analytical technologies as it offers four principal advantages: speed, simplicity of sample preparation, multiplicity of analyses from a single spectrum and the intrinsic non-consumption of the sample (McClure, 1994). In plant breeding programmes, fast, reliable and non-destructive analysis is required in order to quantify valuable components in single plants. For this purpose, NIR spectroscopy has proved to be a very useful tool. During recent years, the growing interest in NIRS has arisen because of improvements in instrumentation and data analysis, as well as the introduction of optical fibres that allow delivery and transfer of NIR energy and information over long distances (Murray and Cowe, 1992). Different types of NIRS analysers are available, ranging from simple filter systems to scanning systems that are able to scan a complete spectrum in a fraction of a second. The first application of Fourier transform to NIR was reported in the 1980s (McClure and Davis, 1988). The extraction of quantitative information from complex NIR spectra has been a notable success of multivariate calibration techniques (Martens and Naes, 1989). Some portable NIR instruments have been developed that promise to be very useful tools in agriculture for on-site measurements as sample preparation is minimal. NIR technology has been applied to quantification of the agricultural commodities of plant and animal origin and their derived products, forages, foods, feed and their ingredients (Williams, 2001). In natural product research, NIR has been used for rapid determination of the active constituents of medicinal plants (Gray *et al.*, 2001; Laasonen *et al.*, 2002; Rager *et al.*, 2002). One of the potential applications of the technology is in quality control, since large lots or single samples can be pre-screened rapidly in order to identify those samples that might require further testing by more time-consuming and expensive methods (Williams, 2001).

Quantitative measurements by NIR are usually based on the correlation between the NIR absorption at different wavelengths and the data of sample composition obtained by the reference method (Table 7.3). The absorption peaks of NIR are broad and overlap, thereby making single wavelength calibration impossible due to large hidden information in the spectral data. Absorption in the NIR arises from the vibrational motion of the molecules. NIR spectra (12,000–4000 cm^{-1} corresponding to the overtone and combination bands arising from the unharmonic nature of the fundamental molecular vibrations of C-H, O-H and N-H groups) are composed of many data points containing significant information about the samples. NIR spectroscopy has been widely used in many fields, including the classification and identification of medicinal plants (Liu *et al.*, 2000; Wu *et al.*, 2000). Some medicinal plants that have a variety of constituents can be identified readily from their NIR spectra according to the position, number and strength of peaks (Liu *et al.*, 2000).

For some medicinal plants, the results have demonstrated that the method could be used as a quick and efficient means of identification with high accuracy. NIRS was used as a rapid and non-destructive analysis tool for curcuminoids in turmeric. Partial least square regression and principal component regression were used as regression methods for NIRS prediction calibration equations, the NIRS spectra and the HPLC reference values of the different curcuminoids. The correlation between the experimental HPLC data and statistical interpretation of the NIRS data distribution was calculated. Performance of the models was evaluated using standard deviation (SD), correlation (R^2), standard error of calibration (SEC), standard error of cross validation (SECV) and standard error of prediction (SEP). The ratio of prediction to deviation (RPD = SD/SEP) was used as an index to check the robustness of the model. A relatively high RPD value indicates a model that can predict chemical composition (Cozzolino *et al.*, 2004). RPD was defined using the following ranges: 0.0–2.3 as not recommended (very poor); 2.4–3.0 as very rough screening (poor); 3.1–4.9 as screening (fair); 5.0–6.4 as quality control (good); 6.5–8.0 as process control (very good); and over 8.1 as suitable for any application (excellent) (Williams, 2001; Fearn, 2002). Nicolai *et al.* (2007) and Pissard *et al.* (2013) have reported that an RPD between 1.5 and 2 means that the model can discriminate low from high values of the response variable; a value between 2 and 2.5 indicates that coarse quantitative predictions are possible and a value between 2.5 and 3 or above corresponds to good and excellent prediction accuracy, respectively.

The spectra of 129 powder samples were similar in shape and all showed strong absorption bands at approximately 1930 nm, which was related mainly to the combination of stretching and deformation of the O-H bonds (water), 2110 nm, which was related to stretching of the N-H bond (protein), and 2310 nm, which was related to the vibration of stretching and deformation of the C-H bond (CH_2). Moreover, the weak absorption band at about 1450 nm was related to the first overtone of the C-H stretching bond (CH_3) and 1765 nm was related to the first overtone of the C-H stretching bond (CH_2) (Shenk *et al.*, 1992).

Seven different absorption bands in the second derivative spectrum with associated curcuminoids were observed at: 1200 nm (C-H stretching second overtone, CH_3); 1440 nm (O-H stretching first overtone, starch and water); 1700 nm (C-H stretching first overtone, CH_2); 1920 nm (C=O stretching second overtone, CONH); 2060 nm (N-H stretching second overtone, $CONH_2$); 2140 nm (C-H stretching plus C=C stretching); and 2280 nm (C-H stretching plus C-H deformation). Raw NIR spectra of turmeric powder samples contained information on the relative characteristics of O-H bonds (found in water), C-H bonds (found in CH_2 and CH_3) and N-H bonds (protein). However, it was difficult to determine due to overtone and combination. Mathematical treatments of the NIRS spectra were required to yield the highest prediction results, and chemometric methods were applied to use information from the spectra to analyse curcuminoids in turmeric samples. A PLS algorithm was used to develop predictive models for curcuminoids. Optimum mathematical treatment of individual curcuminoids resulted in R^2 of 0.943 and SEC of 0.025 for bisdemethoxycurcumin, R^2 of 0.917 and SEC of 0.049 for demethoxycur-cumin, R^2 of 0.929 and SEC of 0.106 for curcumin and R^2 of 0.922 and SEC of 0.178 for total curcuminoids. The cross-validation and external valid-ation statistics for the NIRS model are shown in Table 7.2, which summar-izes the performance parameters computed for the calibration equations. Overall, the coefficients of determination in the cross-validation (R^2) for curcuminoids were 0.927–0.948 and SECV was 0.040–0.201 g g^{-100}. The R^2 values for curcuminoids were 0.854–0.901 and SEP was 0.018–0.129 g g^{-100} in the external validation set. Tanaka *et al.* (2008) first reported the prediction of the presence of curcuminoid using NIRS. Calibration was developed for the ratio of curcuminoids. Kasemsumran *et al.* (2010) reported total curcuminoid not individual curcuminoid.

The feasibility of NIRS application was explored as an alternative to HPLC in the analysis of individual ginsenosides and total ginsenoside content from *Panax quinquefolium* (Reng and Chen, 1999). NIRS has advantages over HPLC with respect to sample preparation and analysis time. Separation of ginsenosides (Rb_1, Rb_2, Rc, Rd, Re, Rg_1, Ro, malonyl-Rb_1, malonyl-Rb_2, malonyl-Rc and malonyl-Rd) in root fibres and main root was performed by HPLC using an RP-18 end-capped column eluted with a phosphate buffer (pH 5.8): acetonitrile gradient with UV detec-tion at 203 nm. NIR spectra were recorded between 400 and 2500 nm. It was concluded that the NIR method could be used for the analysis of the major ginsenosides Rb_1, Re and malonyl-Rb_1, as well as for the total gin-senoside content. NIRS was also used for the determination of total sugar content in Chinese ginseng. Combined with PLS calibration analysis, the NIR spectroscopic method showed the same precision as was obtainable using common chemical methods, indicating that the technique was suit-able for rapid quantitative determination of total sugar content in Chinese ginseng.

Table 7.2. Cross-validation and external validation statistics for curcuminoid content by the NIRS model.

Curcuminoids	Cross-validation set				External validation set			
	PC	R^{2a}	SECV	RPD	PC	R^{2a}	SECV	RPD
Bisdemethoxycurcumin	9	0.948	0.040	2.804	9	0.854	0.018	2.646
Demethoxycurcumin	7	0.943	0.057	3.083	7	0.890	0.040	2.968
Curcumin	8	0.934	0.122	3.258	9	0.901	0.067	3.240
Total curcuminoid	9	0.927	0.201	3.175	8	0.860	0.129	2.715

Notes: SDCV = standard deviation of cross validation, SDP = standard deviation of prediction
PC = principal component; SECV = standard error of cross validation (g 100 g^{-1}); RPD = ratio of prediction to deviation.
[a]R^2 = coefficient of multiple correlations in calibration.
[b]RPD = SDCV/SECV.
[c]RPD = SDP/SECV.

Table 7.3. Wavelength (in nm) and wave numbers (in cm⁻¹) of some NIR bands of organic compounds. (From Stuart, 2004.)

Wavelength (nm)	Wave numbers (cm⁻¹)	Assignment
2,500	4,000	Combination S-H stretching
2,200–2,460	4,545–4,065	Combination C-H stretching
2,000–2,200	5,000–4,545	Combination N-H stretching, combination O-H stretching
1,620–1,800	6,173–5,556	First overtone C-H stretching
1,400–1,600	7,143–6,250	First overtone N-H stretching, first overtone O-H stretching
1,300–1,420	7,692–7,042	Combination C-H stretching
1,100–1,225	9,091–8,163	Second overtone C-H stretching
1,020–1,060	9,804–9,434	Combination S=O stretching
950–1,100	10,526–9.091	Second overtone N-H stretching, second overtone O-H stretching
850–950	11,765–10,526	Third overtone C-H stretching
775–850	12,903–11,765	Third overtone C-H stretching
600–700	1,667–14,286	Combination C-S stretching
450–550	22,222–18,182	Combination S-S stretching

Table 7.4. Some characteristics of NIR and Raman spectroscopies.

	Raman	NIR
Wave number	50–4,000 cm⁻¹	4,000–12,500 cm⁻¹
Bonds	Homonuclear bonds such as C-C, C=C, S-S	H-containing bonds such as C-H, O-H, N-H, S-H
Absorption due to	Scattered radiation	Absorbed radiation (overtones and combination)
Absorption	Strong	Weak
Signal intensity	Poor	Good
Quantification	Intensity versus concentration	Log $(I_0\ I^{-1})$ (Lambert Beer law)
Excitation conditions	Change in polarizability, α	Change in dipole moment, μ
Selectivity	High	Low, requires calibration and chemometrics
Interference	Broad fluorescence baseline	Water, physical attributes (sample size, shape and hardness)
Particle size	Independent	Dependent
Applicability for at-line, online, in-line	Good	Good
Radiation source	Monochromatic (laser VIS/NIR region)	Polychromatic by globar tungsten
Sample preparation	None	None

NIRS was applied for the accurate quantification of chicoric acid in purple coneflower (*Echinacea purpurea*). Root tissues of 169 plants were analysed for chicoric acid by HPLC. Root samples were scanned between 1100 and 2498 nm in reflectance mode. The HPLC data were regressed against infrared spectra to develop empirical prediction equations. The optimum prediction equation produced coefficients of determination for calibration and cross-validation of 0.90 and 0.86, respectively.

A quantification method based on NIRS for total polysaccharides and triterpenoids in *Ganoderma lucidum* and *Ganoderma atrum* from different origins was developed by Chen *et al.* (2012). Spectra were recorded in the 4000–10,000 cm^{-1} regions and four data preprocessing methods (MSC, SNV, first derivative and second derivative) were applied. PLS regression was used in the development of a quantification model for polysaccharides and triterpenoids. The results obtained were comparable with those obtained by the reference method for both components.

Microscopic examination is frequently applied for identification in traditional Chinese medicine (TCM) by looking for the presence of microstructures (Jiang, 2008; Jiang and Chen, 2008; Liu and Li, 2011). Many substances are similar in appearance; therefore, identification by microscopy becomes difficult. Also, identification using microscopy still relies on the experience of the pharmacist, whereas NIRS provides a more reliable, non-subjective method for identification. Liu *et al.* (2006) reported the identification and classification of 269 samples of *Radix Angelicae Dahuricae* and 380 samples of wild or cultivated *Salviae Miltiorrhizae Radix* by NIRS with an accuracy rate of 99 and 95%, respectively.

NIRS was applied to determine the total concentration of phenols in 17 medicinal plants from Romania (Monica *et al.*, 2008). Spectra were recorded in the range 800–2500 nm. Correlation between the values for total phenol content (mM g^{-1}) obtained using Folin–Ciocalteu reagent and predicted values by regression equation on the basis of the reflectance from NIR spectra using the PLS laverage method had a correlation coefficient (R^2) equal to 0.995.

Quantitative determination of naphthodianthrone and phloroglucine derivates in St John's Wort extracts by NIRS was established (Huck *et al.*, 2006b). Three hundred and twenty NIR spectra of 80 extracts were recorded with a scanning polarization interferometer crosswise over a wavelength range from 4500 to 10,000 cm^{-1} in the transflection mode with an optic glass fibre. Ten scans were used for one average spectrum to equilibrate inhomogeneities. Normalization allowed for minimization of baseline shift. The identification of characteristic absorption bands was carried in first derivative spectra. The most intensive bands in the spectrum belonged to vibration of the second overtone of the carbonyl group (5352 cm^{-1}), followed by C-H stretch and C-H deformation vibration, -OH vibration (4440 cm^{-1}) and the CH_2 overtone (5472 cm^{-1}). All 80 extracts were analysed fourfold by liquid chromatography. Validation showed that the robustness of the NIRS model was high, which was demonstrated by the similarity of results for SEE and SEP values: 0.52 and 0.50 µg ml⁻¹ for hypericine and hyperforine, respectively. Accuracy measured in terms of bias (1.6 and 4.2 e-14) was high, and also the correlation coefficients were 0.994 and 0.985, respectively, for hypercine and hyperforine. It was concluded that the model could be used to predict hypercine and hyperforine in liquid extracts of St John's Wort extracts.

The application of NIRS to the simultaneous prediction of alkaloid and phenolic substances in green tea leaves, as well as the discrimination of tea leaves of different ages, was reported (Schulz *et al.*, 1999b). Individual catechins, gallic acid, caffeine and theobromine were determined using reverse phase HPLC. Total polyphenols were determined using colorimetric Folin–Ciocalteu assay. For the prediction of gallic acid, (–)-epicatechin, (–)-epigallocatechin, (–)-epicatechin gallate, (–)-epigallocatechin gallate, caffeine and theobromine, very good calibration statistics were obtained using the partial least squares algorithm (R^2 >0.85), with standard deviation/standard error of cross-validation (SD/SECV) ratio ranging from 2.00 to 6.27. Simultaneously, the dry matter content of the tea leaves could be analysed very precisely (R^2 = 0.94; SD/SECV = 4.12). However, the prediction of total phenol content could be performed with a lower accuracy, which might be due to the lack of specificity in the colorimetric reference method. Using PCA, it was also possible to discriminate tea leaves of different age. Thus, NIRS technology could be applied successfully as a rapid method not only for breeding and cultivation purposes but also to estimate the quality and taste of green tea and to control industrial processes such as decaffeination.

An NIRS method for the rapid quantitative determination of caffeine, theobromine and theophylline in liquid coffee was carried out by recording 249 NIR spectra of 83 samples originating from different geographical regions of the world (Huck *et al.*, 2006a). Spectra in the reflectance or transflectance mode were recorded over a wavelength range from 4009 to 9996 cm⁻¹. For the selection of spectra and wavelength, mathematical pretreatment and statistical analysis, a model was created using chemometrics. The selection of the best regression model was based on the following values calculated for validation purposes: (i) standard error of estimation, standard deviation of the differences between liquid chromatography-ultra violet visible (LC-UV) or liquid chromatography-electrospray ionization-mass spectrometry (LC-ESI-MS) and NIRS results in the calibration set; (ii) standard error of prediction, the counterpart for test samples; (iii) bias, the average deviation between the predicted values and actual values; and (iv) the correlation coefficient (R^2). All recorded spectra were transformed to their first derivative before calculating the cluster model. The most intensive band in the spectrum belonged to vibration of the second overtone of the carbonyl group (5352 cm⁻¹), followed by C-H stretch and C-H deformation vibration (7212 cm⁻¹), -OH vibration of water (4440 cm⁻¹), -CH2 (5742 cm⁻¹) and -CH3 overtones (5308 cm⁻¹). Although the water peak exhibited a significant shift from the maxima of pure water, the discrimination was due mainly to water (6877, 5102 and

4440 cm^{-1}) and lipids (1170, 5797, 5666, 4305 and 4261 cm^{-1}). Vibration of the carbonyl group, -C-H and -CH2 was caused by ingredients such as proteins, lipids, volatile and non-volatile acids, chlorogenic acids, alkaloids and some aroma compounds. Validation of extracts showed that the robustness and reproducibility (BIAS: 5.9 × 10^{-15} for caffeine and theobromine; R^2: 0.86 and 0.85 for caffeine and theobromine, respectively) of the NIRS model was high for the determination of caffeine and theobromine; therefore, the model can be used to predict their content in liquid extracts.

NIRS was applied for the determination and chemical composition of saffron (Zalacain *et al.*, 2005). The validation of 111 samples collected from three main producer contries (Iran, Greece and Spain) was performed with results obtained by UV-visible and HPLC-diode array detector (DAD) measurements. From the validation data, it was demonstrated that this technique was appropriate for determination of the moisture and volatile content and colour strength, and also could be used to estimate the content of five main crocetin glycosides, the compounds responsible for the saffron colour, the best correlations being for trans-crocetin di-(β-D-gentibiosyl) ester (R^2 = 0.93), trans-crocetin di-β-D-glucosyl (R^2 = 0.94) and picrocrocin (R^2 = 0.92), the compound accepted as being responsible for the bitterness of saffron. Further, discriminate analysis among the three geographical regions revealed that the Iranian samples were the most different, whereas the Greek and Spanish samples were more similar.

NIRS can be used to locate the source of a particular sample as variation in weather conditions and soil environment lead to variations in quality (Zhang *et al*, 2011). Qualitative analysis by NIRS not only identified, with 100% accuracy, 22 samples as originating from Henan province (China) but also sourced 68 samples from other provinces correctly and sourced only 9 samples incorrectly (Li *et al.*, 2013).

Samples in the form of powder obtained after grinding loose their significant characteristics, making identification difficult. Using PCA and cluster analysis to classify the NIRS data, it is possible to identify and classify the powder samples as accurately as by HPLC. NIRS is now considered the technique of choice for the quality control of Chinese patent medicines (Xu and Ling, 2010; Chen *et al.*, 2011; Zheng and Han, 2011).

A combination of three-layer, back-propagation ANN with NIR spectra to identify rhubarb, a well-known traditional medicinal material used in China, was applied (Xiang *et al.*, 2002). Pulverized powder samples were scanned using an FT-NIR instrument. The spectra obtained were transferred to first derivative spectra and the principal components were analysed from 519 data points of the derivative spectrum (4000–8000 cm^{-1}). Some of the principal components were selected as the input data for the ANN.

7.2.6 Determination of essential oil constituents

NIRS has high potential to determine the qualitative differences between and simultaneously to quantify the composition of essential oils. Although nowadays GC analysis is very rapid (through the application of fast and ultrafast GC) and may provide competitive and comparable results in 1–3 min, the NIR procedure, however, may be even more rapid, since once the model is created, it requires only a few seconds to analyse the individual oil. Coupled with GC, as the reference technique, NIR technology is a fast and reliable method. The majority of absorption bands in the near infrared spectra of essential oils arise from the overtones of hydrogenic stretching vibrations or combination involving stretching and bending modes (Schulz *et al.*, 1999a). The absorption bands observed at 1135–1215 nm are due to methylene (CH) stretching (2nd (3n) overtone and 2n combination bands). The peaks at 1375–1399 nm are related to methyl (CH) stretch and bending combinations (2nd (3n) overtone and 2n combination bands). The peak around 1690–1695 nm is associated with methyl assymetric stretching (1st (2n) overtone (CH) stretch) (Westad *et al.*, 2008). The intense peaks centred near 2308 and 2348 nm are related to the combination of CH stretching vibration and deformation tones (Hourant *et al.*, 2000).

NIRS was used for the rapid and accurate determination of essential oil content and composition in intact fennel fruits (Steuer and Schulz, 2003) and dried mint leaves (Schulz *et al.*, 1999a) and in the estimation of the different components in isolated essential oils including mint, eucalyptus and citrus (Schulz *et al.*, 1999b; Steuer *et al.*, 2001; Wilson *et al.*, 2001). Leaves and oil samples were scanned with a near infrared monochromator NIR system in the range of 1100–2498 nm to produce a total of 700 points per spectrum with an analysis time of approximately 0.5–1 min. Measurements of dry leaves were performed in the reflectance mode directly on the freshly sieved plant material (approximately 2 g), and the oil samples (approximately 1.5 ml) were analysed in the transmission mode using quartz cuvettes equipped with a diffuse gold reflector (path length: 0.2 mm). The mint oil spectra dominated overtones or different combinations of CH-stretching and bending vibration. The spectral data were treated with weighted multiple scatter correction to eliminate interferences of scatter and transformed with second or third derivative processing. A calibration program was set up with the full wavelength range using the PLS algorithm. The accuracy of the calibration model was described by the multiple coefficients of determination (R^2) and the overall error between the modelled and reference values. The simultaneous prediction of the oil content and its main terpenoid components, menthol and menthonone, could be performed reliably by NIRS measurements. For components occurring in smaller amounts (limonene, 1,8-cineole, menthofuran, isomenthone, pulegone and methyl acetate), a semi-quantitative determination was possible in the dried leaf material. The selection to get mint genotypes with best quality properties (high oil/menthol content and low pulegone content) was reported as the main advantage of the method in the field of applied breeding research. Additionally, there exists the option to characterize reliably and speedily the actual quality parameters during harvest time. Therefore, it is possible to recognize the oil and menthol yields in very short sequences and also to minimize the content of menthofuran, which is undesirable due to the unpleasant aroma in the oil. *Ravensara aromatica* Sonn (Lauraceae, syn *Ravensara anisata*; ravensara) is a well-known endemic aromatic tree from Madagascar. All its aerial parts contain essential oil comprising limonene (14–23%) and sabinene (10–16%) with methyl chavicol (2–12%). Its bark, reddish in colour, is rich in methyl chavicol (90–95%). Essential oil (1 ml) was scanned in the range of 300–2500 nm using a portable NIR spectrometer. The major components of the essential oil, namely, 1,8-cineole (58%), sabinene (11%) and α-terpineol (8%), were predicted accurately by the cross-validation model with a high coefficient of determination ($r^2 \geq 0.993$), and the variance between the modelled and the reference values was less than 1%.

Wilson *et al.* (2001) proposed that NIRS could be used as an alternative method for the determination of cineole content in eucalyptus oils. NIR spectroscopy was applied for the estimation of 1,8-cineole (eucalyptol) in eucalyptus oil. Thirty different eucalyptus oil samples were scanned in reflectance mode. The cineole content of each sample was determined by the British Pharmacopoeia (BP) monograph method, and these reference data were used to construct two calibration equations for the cineole content in the oils. The mean accuracy for the NIR method differed by 1.01% or less and the mean bias by ±0.33% or less, compared with the BP method. Calculation of the slope and intercept of plots of NIR predicted values at 95% confidence intervals against the BP method reference values showed that there was no evidence of fixed or relative systematic errors. The range of cineole contents used in the calibrations was further extended by incorporating five samples of eucalyptus oil spiked with cineole and five samples of two essential oils known to have lower cineole content than eucalyptus oil, to give a range of 52.5–99.0% (w/w). The mean accuracy decreased to an error of 1.26% or less and the bias to ±0.50% or less. Again, confidence intervals suggested there was no evidence of fixed or systematic errors in the NIR calibrations.

Lemongrass oil and lemon oil contains 65–85% (w/w) and 2–5% citral (w/w), respectively. Wilson *et al.* (2002) reported accurate and approximate determination of citral in lemongrass oil and lemon oil, respectively, using NIR spectroscopy. A total of 26 samples of pure lemongrass oil and 35 samples of pure lemon oil (both including samples that were spiked with citral to increase the calibration range) were scanned using a reflector vessel as the sample presentation method. The calibration data were generated using the British Pharmacopoeia (BP) monograph method. Mean accuracy was found to be 1.00 or less and the mean bias was 0.09% or less for lemongrass. For lemon oil, the mean accuracy was found to be 4.28% or less and the mean bias was 0.71% or less.

Kuriakose *et al.* (2010) reported the application of NIRS to detect adulterants (even 1% of the low-grade oils) in high-quality sandalwood oil. After suitable preprocessing of the NIR raw spectral data, the models were built up by cross-validation and calibration (RMSECV and RMSEC, % v/v) as a decision-supporting system to fix the optimal number of factors.

The coefficients of determination (R^2) and the root mean square error of prediction (RMSEP, % v/v) in the prediction sets were used as the evaluation parameters ($R^2 = 0.9999$ and RMSEP = 0.01355).

Kuriakose and Joe (2012) reported the successful application of NIRS assisted by multivariate chemometric techniques such as PCA, a pattern recognition technique, namely hierarchical cluster analysis (HCA) (Armenta *et al.*, 2007), and a self-organizing map (SOM) (Kohonen *et al.*, 1996) to the classification of sandalwood oils according to their quality. Quantification of the constituents, especially adulterants, if any, in sandalwood oils was also achieved by support vector machine regression (SVMR), used to quantify the percentage level of counterfits in the oils (Vapnik, 1995) with proper combinations of data pretreatments.

Steuer *et al.* (2001) reported the application of NIRS in the classification and analysis of citrus oils. Citrus oils are used to flavour sweets and juices, as well as in applications that demand a fresh odour. Limonene is the main component of citric peel oil. Citrus oil also contains different aldehydes such as octanal, decanal and citral, as well as sinesal, which contribute very strongly to the whole aroma impression. Due to the formation of artefacts resulting from exposure to acid and air, distilled oils may contain content especially of *p*-cymene, terpinen-4-ol and α-terpineol in comparison to the related cold-pressed oil. Citrus oil samples were scanned in the range from 1100 to 2500 nm with a resolution of 2 mm in trans-reflectance mode using quartz cuvettes, and PCA was performed for the spectral data. Calibration was developed using a modified partial least squares algorithm with an optimum number of PLS factors. The calibration accuracy was described by the multiple coefficient of determination (R^2) and overall error between the modelled and reference values (SECV).

Essential oil and its composition in the intact fruits of caraway (*Carum carvi*) and fennel (*Foeniculum vulgare*) was determined by dispersive and FT-NIR spectrometers (Schulz *et al.*, 2000). Spectral data were pretreated with a multiple scatter correction and then transformed with second or third derivative processing. An MPLS algorithm was used on the spectra and reference values for the prediction of the content and composition of the individual oil by dispersive NIR spectrometer. For the FT-NIR spectrometer, the optimum number of PLS ranks was used for the respective predictions. For statistical parameters such as R^2 and SECV, only small differences between the results of dispersive and Fourier transform measurements were found. Sampling schemes were observed as the major concern than the optical configuration.

7.3   Raman Spectroscopy

Fourier transform Raman spectroscopy has been described as a successful non-destructive technique employed in analytical investigations of natural products (Edwards *et al.*, 2005a,b). Raman spectroscopy is a very promising tool in analytical chemistry as many samples can be examined non-destructively in a short time with no sample preparation. Moreover, Raman spectra exhibit well-resolved bands of fundamental vibrational transitions, thus providing a high content of molecular structural information (Muik *et al.*, 2003b; Table 7.4). In combination with chemometric data evaluation, Raman spectrometry is a powerful tool capable of extracting quantitative chemical information from complex matrix (Himmelsbach *et al.*, 2001; Schulz *et al.*, 2002). Recent advances in instrumentation technology have contributed decisively to a rapid increase in the industrial applications of Raman spectroscopy (Chalmers and Dent, 1997; Cooper, 1999).

Raman spectra of raw materials and solvent extracts of *Cacao* seeds were recorded using the macroscopic mode with a specimen footprint of about 100 μm using a Bruker IFS 66 instrument with an FRA 106 Raman module attachment and an Nd^{3+}/YAG laser operating at 1064 nm in the near infrared and an InGaAs detector cooled with liquid nitrogen. The spectrum consisted of several C-H stretching vibrations at 3000–2850 cm^{-1} and some sharp bands that could be considered as key features of the kernel and fat matrix, including C=O stretching bands at 1743 and 1730 cm^{-1} (shoulder), (C=C) + (C=O) stretching at 1659 cm^{-1} and a broad unresolved aromatic C-CH quadrant stretching mode feature centred at 1609 cm^{-1}. The identification of biomarker bands for theobromine, a member of the caffeine group of alkaloids, was done successfully.

Raman spectroscopy was used to investigate polyacetylenes in American ginseng roots (Baranska *et al.*, 2006). The Raman spectra taken *in situ* from fresh ginseng root revealed a characteristic polyacetylene band at 2237 nm^{-1}, whereas in the spectrum obtained from the dried root, this band was shifted to about 2258 nm^{-1}. The data obtained had a good agreement with the spectra obtained using isolated standards (falcarinol 2258 nm^{-1} and panaxydol 2260 nm^{-1}). Further, application of the Raman mapping technique to ginseng roots showed that the content of both main polyacetylenes decreased with increased root size. This finding was confirmed by HPLC analysis.

Raman spectroscopic studies of Guarana, an important product of the Amazon rainforest, was investigated using Raman spectroscopy (Edwards *et al.*, 2005a). The therapeutic properties of guarana and its extracts have been attributed to guaranine, which could be a complex of caffeine and tannins, or to a xanthine natural product. Comparison of the Raman spectra of pure and anhydrous caffeine and of guarana methanolic extracts showed that the extracts of guarana contained anhydrous caffeine, confirmed by the presence of the doublet at 1698 and 1654 cm^{-1}. Also, the composition of guarana was very similar for the whole seed and for the outer and inner portions of the dissected seed. The above findings indicated that Fourier transform Raman spectroscopy could be used to monitor quality control of guarana products in the phytopharmaceutical industry.

Raman spectrometry has been applied sucessfully in determining the total unsaturation in oils, the classification of oil and fat and for the detection of adulterants in virgin olive oil (Sadeghi-Jorabchi *et al.*, 1990; Baeten and Meurens, 1996; Marigheto *et al.*, 1998; Barthus and Poppi, 2001). Fourier transform Raman spectroscopy in combination with PLS regression was used for the direct, reagent-free determination of free fatty acid content in olive oil and olives (Muik *et al.*, 2003a). Oils were investigated directly in a simple flow cell. Milled olives were measured in a dedicated sample cup, which was rotated eccentrically to the horizontal laser beam during spectrum acquisition to compensate sample heterogeneity. The samples were illuminated by a Nd:YAG laser line at 1064 nm with a power of 500 mW using a focused laser beam. Both external and internal (leave-one-out) validation was used to assess the ability of PLS calibration models to predict the free fatty acid (FFA) content (in terms of oleic acid) in oil and olives in the range 0.24–6.14 and 0.15–3.79%, respectively. The root mean square error of prediction (RMSEP) was 0.29% for oil and 0.26% for olives. Ninety per cent of the oil samples and 80% of the olives were classified correctly.

7.4 Biosensors

Biosensors have also found application in phytochemical analysis. Biosensors consist of a biological element that has a direct contact with a transducer. Biological components are immobilized on the surface of the biological recognition element, where they react with the target compounds. The transducers convert a particular biochemical phenomenon such as electrochemical, mass, optical or temperature change into electrical signals (Angelova *et al.*, 2008). A superoxide dismutase biosensor has been developed to evaluate the antioxidant capacity of ginseng tea (Campanella *et al.*, 2004). An optical biosensor based on optical fibres and laser technology has been developed to distinguish ginseng from sawdust (Yap *et al.*, 2005). Spectroscopic analysis was performed directly on powdered root samples. It was found that a PCA of the fingerprint region between 2000 and 600 cm^{-1} could be applied to distinguish between ginseng and sawdust, as well as between ginsengs.

7.5 Conclusion

NIRS belongs to indirect or secondary methods. To obtain information, sample spectra have to be compared to the primary or reference methods. Mathematical means are applied to spectra to establish an NIR predictive model to predict the composition of the investigated samples.

NIRS technology is developing and improving continuously. Its wider application for the quantitative and qualitative analysis of natural products will certainly improve their quality control and safety in clinical use. The further development of NIRS will serve to strengthen quality monitoring and control of natural products and regulate the market. Both solid and liquid samples in different types of packaging can be tested because of the better penetrability of the fibre optics used in NIRS (Xing and Zhang, 2010). NIRS is a full-spectrum method in constrast to chromatographic or electrophoretic analytical methods, which focus only on the separation and detection of certain analytes. Further, the NIRS method not only allows offline but also online or even in-line analysis. Even though the costs of NIRS equipment are high and calibration needs a lot of time, NIRS has the great advantage of reducing time and costs, especially in combination with the fast reference method.

Since, in principle, NIRS measurements can also be performed on fresh plant material, it is possible to use this method to predict the optimal time to harvest or for the selection of individual genotypes directly in field. Continuing developments in diode array instruments are expected to extend the potential of NIR for on-site determination to various plant constituents.

Non-destructive techniques could be utilized as a control procedure or as an alternative rapid and effective quantification method. These analytical tools perform as well as established conventional methods, while reducing the cost of analysis in terms of chemical reagent use, which is more environmental friendly. The following points are worthy to mention for the proper application of NIRS:

1. Prior to NIRS analysis, reference methods such as GC, HPLC, etc., have to be used. HPLC has to be used in order to build a reference data set, as the NIRS method was developed using a statistical comparison of the results obtained from the reference data.

2. Samples are divided randomly into two groups: calibration set (approximately 66% of samples) and validation set (approximately 34% of samples).

3. A wide range of analyte concentration is helpful to establish a stable and accurate calibration model.

4. Calibration of the lower SEC as well as higher R^2 values is considered more appropriate and much more accurate (Font *et al.*, 2005).

5. A lower SEC and higher SEC values are considered better and more accurate for calibration.

6. The development and maintenance of NIR calibrations is known to be expensive in terms of time and personnel. Therefore, it is useful to perform central calibrations and to transfer them to individual instruments at different locations. There are a number of factors that can influence the success of spectral transfer between diode array instruments and monochromators, such as resolution, wavelength range, absorption density, standardization methods and software options.

7. The NIRS technique has the potential to replace the existing quality control standard methods, which are mostly very time-consuming.

8. NIRS measurements using optical fibres have the advantage of online control during distillation, extraction or other industrial processes.

9. The NIRS technique has high potential to determine the qualitative differences in and simultaneously to quantify the composition of essential oils.

References

Angelova, N., Kong, H.W., Heijden, R.V., Yang, S.Y., Choi, Y.H., Kim, H.K., Wang, M., Hankemeier, T., Greef, J.V., Xu, G. and Verpoorte, R. (2008) Recent methodology in the phytochemical analysis of Ginseng. *Phytochemical Analysis* 19, 2–16.

Armenta, S., Garrigues, S. and Guardia, M.I. (2007) Determination of edible oil parameters by near infrared spectrometry. *Analytica Chimica Acta* 596, 330–337.

Baeten, V. and Meurens, M. (1996) Detection of virgin olive oil adulteration by Fourier transform Raman spectroscopy. *Journal of Agricultural and Food Chemistry* 44, 2225–2230.

Baranska, M., Schulz, H. and Christensen, L.P. (2006) Structural changes of polyacetylenes in American ginseng root can be observed *in situ* using Raman spectroscopy. *Journal of Agriculural and Food Chemistry* 54, 3629–3635.

Barnes, R.J., Dhanoa, M.S. and Lister, S.J. (1989) Standard normal variate transformation and detrending of near infra diffuse reflectance spectra. *Applied Spectroscopy* 43, 772–777.

Barthus, R.C. and Poppi, R.J. (2001) Determination of the total unsaturation in vegetable oils by Fourier transform Raman spectroscopy and multivariate calibration. *Vibrational Spectroscopy* 26, 99–105.

Bhatty, R.S. (1991) Measurement of oil in whole flaxseed by near-infrared reflectance spectroscopy. *Journal of American Oil Chemists Society* 68, 34–38.

Cadet, F. and de la Guardia, M. (2001) Infrared quantitative analysis. In: Meyers, R.A. (ed.) *Encyclopedia of Analytical Chemistry*. John Wiley & Sons Inc., Chichester, UK, pp. 1–26.

Campanella, L., Bonanni, A., Einotti, E. and Tomassetti, M. (2004) Biosensors for determination of total and natural antioxidant capacity of red and white wines: comparison with other spectrophotometric and fluorimetric methods. *Biosensors and Bioelectronics* 19, 641–651.

Centner, V., Massart, D.L., de Jong, S., Vandeginste, B.M. and Sterna, C. (1996) Lamination of uninformative variables for multivariate calibration. *Analytical Chemistry* 68, 3851–3858.

Chalmers, J.M. and Dent, G. (eds) (1997) *Industrial Analysis with Vibrational Spectroscopy*. Royal Society of Chemistry, Cambridge, UK.

Chen, J.Q., Li, Q.H. and Li, Q. (2011) The study of NIR detection on tindantongluo capsules. *Chinese Pharmaceutical Affairs* 25, 36–38.

Chen, Q.S., Zhao, J.W., Zhang, H.D. and Wang, X.Y. (2006) Feasibility study on qualitative and quantitative analysis in tea by near infrared spectroscopy with multivariate calibration. *Analytica Chimica Acta* 572, 78–81.

Chen, Y., Xie, M., Zhang, H., Wang, Y., Nie, S. and Chang, L. (2012) Quantification of total polysaccharides and triterpenoids in *Ganoderma lucidum* and *Ganoderma atrum* by near infrared spectroscopy and chemometrics. *Food Chemistry* 135, 268–275.

Clark, C.J., McGlone, V.A. and Jordan, R.B. (2003) Detection of brown heart in Braeburn apple by transmission NIR spectroscopy. *Post Harvest Biology and Technology* 28, 87–96.

Cooper, J.B. (1999) Chemometric analysis of Raman spectroscopic data for process control applications. *Chemometrics and Intelligent Laboratory Systems* 46, 231–247.

Cozzolino, D., Kwiatkowski, M.J., Parker, M., Cynkar, W.U., Dambergs, R.G., Gishen, M. and Herderich, M.J. (2004) Prediction of phenolic compounds inreds wine fermentations by visible and near infrared spectroscopy. *Analytica Chimica Acta* 513, 73–80.

Cubeddu, R., Pfifferi, P., Taroni, P., Valentini, G., Torricelli, A., Valero, C., Ruiz-Altisent, M. and Ortiz, C. (2001) Nondestructive quantification of chemical and physical properties of fruits by time-resolved reflectance spectroscopy in the wavelength range 650-1000 nm. *Applied Optics* 40, 538–543.

Davies, A.M. and Grant, A. (1987) Review: near infrared analysis of food. *International Journal of Food Science and Technology* 22, 191–207.

Edwards, H.G.M., Villar, S.E.J., de Oliveira, L.F.C. and Hyaric, M.L. (2005a) Analytical Raman spectroscopic study of Cacao seeds and their chemical extracts. *Analytica Chimica Acta* 538, 175–180.

Edwards, H.G.M., Farwell, D.W., de Oliveira, L.F.C., Alia, J.M., Hyaric, M.L. and de Amedida, M.V. (2005b) FT-Raman spectroscopic studies of guarana and some extracts. *Analytica Chimica Acta* 532, 177–186.

Elfadl, E., Reinbrecht, C. and Claupein, W. (2010) Development of near infrared reflectance spectroscopy (NIRS) calibration model for estimation of oil content in a worldwide safflower germplasm collection. *International Journal of Plant Protection* 4, 259–270.

Fearn, T. (2001) Standardisation and calibration transfer for near infrared instruments: a review. *Journal of Infrared Spectroscopy* 9, 229–244.

Fearn, T. (2002) Assessing calibrations, SEP, RPD, RER and R2. *NIR News* 13, 12–14.

Font, R., Rio-Celestino, M., Cartea, E. and de Haro-Bailon, A. (2005) Quantification of glucosinolates in leaves of leaf rape (*Brassica napus* ssp. *pabularia*) by near infrared spectroscopy. *Phytochemistry* 66, 175–185.

Fu, H.Y., Huang, D.C., Yang, T.M., She, Y.B. and Zhang, H. (2013) Rapid recognition of Chinese herbal pieces of areca catechu by different concocted processes using Fourier transform mid-infrared and near infrared spectroscopy combined with partial least squares discriminant analysis. *Chinese Chemical Letters* 24, 639–642.

Fuller, M.P. and Griffiths, P.R. (1978) Diffuse reflectance measurements by infrared Fourier transform spectrometry. *Analytical Chemistry* 50, 1906–1910.

Galtier, O., Dupuy, N., Le Dréau, Y., Ollivier, D., Pinatel, C., Kister, J. and Artaud, J. (2007) Geographic origins and compositions of virgin oils determined by chemometric analysis of NIR spectra. *Analytica Chimica Acta* 595, 136–144.

Gao, H.B., Xiang, B.R., Li, R. and Liu, H. (2006) Rapid, non destructive quantitative analysis of nicotinic acid tablets by NIR combined with wavelet transform and artificial neural network. *Journal of China Pharma University* 37, 326–329.

Geladi, P. and Dabakk, E. (1995) An overview of chemometrics applications in near infrarefred spectrometry. *Journal of Near-infrared Spectroscopy* 3, 119–132.

Gray, D.E., Roberts, C.A., Rottinghaus, G.E., Garrett, H.E. and Pallardy, S.G. (2001) Quantification of root chicoric acid in purple coneflower by near infrared reflectance spectroscopy. *Crop Science* 41, 1159–1161.

Greensill, C.V. and Walsh, K.B. (2000) A remote acceptance probe and illumination configuration for spectral assessment of internal attributes of intact fruit. *Measurement Science and Technology* 11, 1674–1684.

Guillen, M.D. and Cabo, N. (1997) Infrared spectroscopy in the study of edible oils and fats. *Journal of the Science of Food and Agriculture* 75, 1–11.

Guo, L.P., Huang, L.Q. and Huck, C. (2009) Near infrared spectroscopy (NIRS) technology and its application in geoherbs. *Journal of Chinese Materia Medica* 34, 1751–1757.

He, Y., Li, X.L. and Deng, X.F. (2007) Discrimination of varieties of tea using near infrared spectroscopy by principal component analysis and BP mode. *Journal of Food Engineering* 79, 1238–1242.

Himmelsbach, D.S., Barton, F.E. II, McClung, A.M. and Champagne, E.T. (2001) Protein and apparent amylose contents of milled rice by NIR-FT/Raman Spectroscopy. *Cereal Chemistry* 78, 488–492.

Hou, J., Chen, G.S. and Wang, Z.P. (2001) Development of artificial neural network and its application in multivariate calibration. *Journal of Analytical Science* 17, 68–74.

Hourant, P., Baeten, V., Morales, M.T., Meurens, M. and Aparico, R. (2000) Oil and fat classification by selected bands of near infrared spectroscopy. *Applied Spectroscopy* 54, 1168–1174.

Huck, C.W., Guggenbichler, W. and Bonn, G.K. (2006a) Analysis of caffeine, theobromine and theophylline in coffee by near infrared spectroscopy (NIRS) compared to high-performance liquid chromatography (HPLC) coupled to mass spectrometry. *Analytica Chimica Acta* 538, 195–203.

Huck, C.W., Abel, G., Popp, M. and Bonn, G.K. (2006b) Comparative analysis of naphthodian-throne and phloroglucine derivatives in St. John's Wort extracts by near infrared spectroscopy, high performance liquid chromatography and capillary electrophoresis. *Analytica Chimica Acta* 580, 223–230.

Jiang, H. (2008) The identifying TCM-crop by hemi micro technology. *Chinese Journal of Ethnomedicine and Ethnopharmacology* 4, 54–56.

Jiang, H. and Chen, X. (2008) The quickly identifying TCM-crop by hemi micro shape and properties. *Chinese Journal of Ethnomedicine and Ethnopharmacology* 10, 10–12.

Jolliffe, I.T. (1986) *Principal Component Analysis*. Springer Verlag, New York.

Kamal-Eldin, A. and Andersson, R. (1997) A multivariate study of the correlation between tocopherol content and fatty acid composition in vegetable oils. *Journal of American Oil Chemists Society* 74, 375–380.

Kasemsumran, S., Keeratinijakal, V., Thanapase, W. and Ozaki, Y. (2010) Near infrared quantitative analysis of total curcuminoids in rhizomes of *Curcuma longa* by moving window partial least squares regression. *Journal of Infrared Spectroscopy* 18, 263–269.

Kohonen, T., Oja, E., Simula, O., Visa, A. and Kangas, J. (1996) Engineering applications of the self-organizing map. *Proceedings of the IEE* 84, 1358–1384.

Kumar, S. and Andy, A. (2013) Fourier transform-near infrared reflectance spectroscopy calibration development for screening of oil content of intact safflower seeds. *International Food Research Journal* 20, 759–762.

Kuriakose, S. and Joe, H. (2012) Qualitative and quantitative analysis in sandalwood oils using near infrared spectroscopy combined with chemometric techniques. *Food Chemistry* 135, 213–218.

Kuriakose, S., Thankappan, X., Joe, H. and Venkataraman, V. (2010) Detection and quantification of adulteration in sandalwood oil through near infrared spectroscopy. *Analyst* 135, 2676–2681.

Laasonen, M., Harmia-Pulkkinen, T., Simard, C.L., Michiels, E., Rasanen, M. and Vuorela, H. (2002) Fast identification of *Echinacea purpurea* dried roots using near infrared spectroscopy. *Analytical Chemistry* 74, 2493–2499.

Li, B.X., Wei, Y.H., Duan, H.G., Xi, L.L. and Wu, X.A. (2012) Discrimination of the geographical origin of *Codonopsis pilosula* using near infrared diffuse reflection spectroscopy coupled with random forests and k-nearest neighbor methods. *Vibrational Spectroscopy* 62, 17–22.

Li, W., Chen, J., Xiang, B.R. and An, D. (2000) Simultaneous on-line dissolution monitoring of multicomponent solid preparations containing vitamins B1, B2 and B6 by fibre optic sensor system. *Analytica Chimica Acta* 408, 39–47.

Li, W.H., Cheng, Z.W., Wang, Y.F. and Qu, H.B. (2013) Quality control of *Lonicerae japonicae* Flos using near infrared spectroscopy and chemometrics. *Journal of Pharmaceutical and Biomedical Analysis* 72, 33–39.

Li, Y., Brown, C.W. and Lo, S. (1999) Near infrared spectroscopic determination of alcohols: solving non linearity with linear and non linear methods. *Journal of Near-infrared Spectroscopy* 7, 55–62.

Li, Y.Z., Min, S.G. and Liu, X. (2007) Study on the methods and applications of near infrared spectroscopy chemical pattern recognition. *Spectroscopy and Spectral Analysis* 27, 1299–1303.

Liang, Y.Z. and Kvalheim, O.M. (1994) Diagnosis and resolution of multi wavelength chromatograms by rank map, orthogonal projections and sequential rank analysis. *Analytica Chimica Acta* 292, 5–15.

Liu, G.L., Chai, J.N., Li, W., Wang, Z.T., Zou, Q.G., Xu, L.S. and Xiang, L. (2000) Near infrared spectroscopy technique used in the classification of the *Cnidium monnieri* (L.). Cusson. *Computers and Applied Chemistry* 1, 109–110.

Liu, M.M. and Li, F. (2011) Microscopic identification of new technologies and applications in progress in the identification of Chinese crude drugs. *Journal of Liaoning University of Traditional Chinese Medicine* 13, 50–52.

Liu, M.Q., Zhou, D.C, Xu, X.Y., Sun, Y.J., Zhou, X.L. and Han, L. (2007) Clustering analysis applied to near infrared spectroscopy analysis of Chinese traditional medicine. *Spectroscopy and Spectral Analysis* 27, 1985–1988.

Liu, S.H., Zhang, X.G., Zhou, Q. and Sun, S.Q. (2006) Determination of geographical origins of Chinese medical herbs by NIR and pattern recognition. *Spectroscopy and Spectral Analysis* 26, 629–632.

Lu, H.Y., Wang, S.S., Cai, R., Meng, Y., Xie, X. and Zhao, W.J. (2012) Rapid discrimination and quantification of alkaloids in corydalis tuber by near infrared spectroscopy. *Journal of Pharmaceutical and Biomedical Analysis* 59, 44–49.

Lu, R. (2003) Detection of bruises on apples using near infra hyperspectral scattering. *Transactions in American Society of Agricultural Engineers* 46, 523–530.

Lu, R. and Ariana, D. (2002) A near infrared sensing technique for measuring internal quality of apple fruit. *Transactions in American Society of Agricultural Engineers* 18, 1089–1094.

Lucio-Gutierrez, J.R., Coello, J. and Maspoch, S. (2011) Application of near infrared spectral fingerprinting and pattern recognition techniques for fast identification of *Eleutherococcus senticosus*. *Food Research International* 44, 557–565.

Marigheto, N.A., Kemsley, E.K., Defernez, M. and Wilson, R.H. (1998) A comparison of mid-infrared and raman spectroscopies for the authentication of edible oils. *Journal of American Oil Chemists Society* 75, 987–992.

Martens, H. and Naes, T. (1989) *Multivariate Calibration*. Wiley, Chichester, UK.

Martinsen, P. and Schaare, P. (1998) Measuring soluble solids distribution in kiwifruit using near infrared imaging spectroscopy. *Post Harvest Biology and Technology* 14, 271–281.

McClure, W.F. (1994) Near-infrared spectroscopy – the giant is running strong. *Analytical Chemistry* 66, 43A–53A.

McClure, W.F. and Davis, A.M.C. (1988) Fast Fourier transforms in the analysis of near infrared spectra. In: Creaser, C.S. and Davies, A.M.C. (eds) *Analytical Applications of Spectroscopy*. Royal Society of Chemistry, London, pp. 414–436.

Miller, J.N. and Miller, J.C. (2005) *Statistics and Chemometrics for Analytical Chemistry*, 5th Edition. Pearson Education Limited, Edinburgh Gate, Harlow, UK, pp. 213–239.

Monica, H., Moisuc, A., Radu, F., Dragan, S. and Gergen, I. (2008) Total polyphenol content determination in complex matrix of medicinal plants from Romania by NIR spectroscopy. *Bulletin UASVM, Agriculture* 65, 123–128.

Moschner, C.R. and Biskupek-Korell, B. (2006) Estimating the content of free fatty acids in high oleic sunflower seeds by near infrared spectroscopy. *European Journal of Lipid Science and Technology* 108, 606–613.

Muik, B., Lendl, B., Molina-Diaz, A. and Ayora-Canada, M.J. (2003a) Direct, reagent-free determination of free fatty acid content in olive oil and olives by Fourier transform Raman spectroscopy. *Analytica Chimica Acta* 487, 211–220.

Muik, B., Lendl, B., Molina-Diaz, A. and Ayora-Canada, M.J. (2003b) Fourier transform Raman spectrometry for the quantitative analysis of oil content and humidity in olive. *Applied Spectroscopy* 57, 233.

Murray, I. and Cowe, I.A. (1992) *Making Light Work: Advances in Near Infrared Spectroscopy*. Wiley, Chichester, UK.

Naes, T., Isaksson, T., Fearn, T. and Davies, T. (2004) *A User-friendly Guide to Multivariate Calibration and Classification*. NIR Publications, Charlton, Chichester, UK.

Ni, L.J., Zhang, L.G., Hou, J., Shi, W.Z. and Guo, M.L. (2009) A strategy for evaluating antipyretic efficacy of Chinese herbal medicines based on UV spectral fingerprints. *Journal of Ethnopharmacology* 124, 79–86.

Nicolai, B.M., Beullens, K., Bobelyn, E., Peirs, A., Saeys, W., Theron, K.L. and Lammertyn, J. (2007) Nondestructive measurement of fruit and vegetable quality by means of NIR spectroscopy: a review. *Postharvest Biology and Technology* 46, 99–118.

Norris, K.H. (1964) Design and development of a new moisture meter. *Agricultural Engineering* 45, 370–372.

Pang, H.H., Feng, Y.C., Hu, C.Q. and Xinag, B.R. (2008) Construction of universal quantitative models for determination of cefoperazone sodium for injection from different manufacturers using near infrared reflectance spectroscopy. *Journal of Chinese Pharmaceutical Science* 26, 2214–2218.

Pedro, A.M.K. and Ferreira, M.M.C. (2005) Nondestructive determination of solids and carotenoids in tomato products by near infrared reflectance spectroscopy and multivariate calibration. *Analytical Chemistry* 77, 2505–2511.

Peirs, A., Ooms, K., Lammertyn, J. and Nicolai, B. (2002a) Comparison of Fourier transform and dispersive near infrared reflectance spectroscopy for apple quality measurements. *Biosystem Engineering* 81, 305–311.

Peirs, A., Tirry, J., Verlinden, B., Darius, P. and Nicolai, B. (2002b) Effect of biological variability on the robustness of NIR models for soluble solids content of apples. *Post Harvest Biology and Technology* 28, 269–280.

Pissard, A., Pierena, J.A.F., Baeten, V., Sinnaeve, G., Lognay, G., Mouteau, A., Dupont, P., Rondia, A. and Lateur, M. (2013) Non-destructive measurement of vitamin C, total polyphenol and sugar content in apples using near-infrared spectroscopy. *Journal of Science of Food and Agriculture* 93, 238–244.

Qi, X.M., Zhang, L.D., Du, X.L., Song, Z.J., Zhang, Y. and Xu, S.Y. (2003) Quantitative analysis using NIR by building PLS-BP model. *Spectroscopy and Spectral Analysis* 23, 870–872.

Rager, I., Ross, G., Schmidt, P.C. and Kovar, K.A. (2002) Rapid quantification of constituents in St. John's Wort extracts by NIR spectroscopy. *Journal of Pharmaceutical and Biomedical Analysis* 28, 439–446.

Reng, G.X. and Chen, F. (1999) Simultaneous quantification of ginsenosides in American ginseng (*Panax quinquefolium*) root powder by visible/near infrared reflectance spectroscopy. *Journal of Agricultural Food Chemistry* 47, 2771–2775.

Roggo, Y., Duponchel, L., Ruckebusch, C. and Huvenne, J.P. (2003) Statistical tests for comparison of quantitative and qualitative models developed with near infrared spectral data. *Journal of Molecular Structure* 654, 253–262.

Sadeghi-Jorabchi, P.J., Hendra, R.H., Wilson, R.H. and Belton, P.S. (1990) Determination of the total unsaturation in oils and margarines by Fourier transform Raman spectroscopy. *Journal of American Oil Chemists Society* 67, 483–486.

Schulz, H., Drews, H.H. and Kruger, H. (1999a) Rapid NIRS determination of quality parameters in leaves and isolated essential oil of *Mentha* species. *Journal of Essential Oil Research* 11, 185–190.

Schulz, H., Engelhardt, U.H., Wegent, A., Drews, H.H. and Lapczynski, S. (1999b) Application of near infrared reflectance spectroscopy to the simultaneous prediction of alkaloids and phenolic substance in green tea leaves. *Journal of Agricultural and Food Chemistry* 47, 5064–5067.

Schulz, H., Quilitzsch, R., Drews, H.H. and Kruger, H. (2000) Estimation of minor components in Caraway, Fennel and Carrots by NIRS comparison of results from dispersive and Fourier-transform instruments. *International Agrophysics* 14, 249–253.

Schulz, H., Schrader, B., Quilitzsch, R. and Steuer, B. (2002) Quantitative analysis of various citrus oils by ATR/FT-IR and NIR-FT Raman spectroscopy. *Applied Spectroscopy* 56, 117.

Shenk, J.S. and Westerhaus, M.O. (1991) New standardization and calibration procedures for NIRS analytical systems. *Crop Science* 31, 1694–1696.

Shenk, J. and Westerhaus, M.O. (1993) *Analysis of Agriculture and Food Products by Near-infrared Reflectance Spectroscopy*. Infrasoft International, Port Matilda, Pennsylvania.

Shenk, J., Westerhaus, M.O. and Workman, J.J. (1992) Application of NIR spectroscopy to agricultural products. In: Burns, D. and Ciurczak, E. (eds) *Handbook of Near-Infra Analysis*. Marcel Dekker, New York, pp. 383–431.

Sinelli, N., Spinardi, A., Egidio, V.D., Mignani, I. and Casiraghi, E. (2008) Evaluation of quality and nutraceutical content of blueberries (*Vaccinium corymbosum*) by near and mid-infrared spectroscopy. *Postharvest Biology and Technology* 50, 31–36.

Steuer, B. and Schulz, H. (2003) Near infrared analysis of Fennel (*Foeniculum vulgare* Miller) on different spectrometers – basic consideration for a reliable network. *Phytochemical Analysis* 14, 285–289.

Steuer, B., Schulz, H. and Lager, E. (2001) Classification and analysis of citrus oils by NIR spectroscopy. *Food Chemistry* 72, 113–117.

Stratis, D.N., Eland, K.L., Carter, J.C., Tomlinson, S.J. and Angel, S.M. (2001) Comparison of acousto-optic and liquid crystal tunable filters for laser induced breakdown spectroscopy. *Applied Spectroscopy* 55, 999–1004.

Stuart, B.H. (2004) *Infrared Spectroscopy: Fundamentals and Applications*. John Wiley & Sons, Ltd, Chichester, UK.

Tanaka, K., Kuba, Y., Sasaki, T., Hiwatashi, F. and Komatsu, K. (2008) Quantitation of curcuminoids in curcuma rhizome by near infrared spectroscopic analysis. *Journal of Agricultural and Food Chemistry* 56, 8787–8792.

Tian, R.T., Xie, P.S. and Liu, H.P. (2009) Evaluation of traditional Chinese herbal medicine: chaihu (Bupleuri Radix) by both high performance liquid chromatographic and high performance thin layer chromatographic finger print and chemometric analysis. *Journal of Chromatography A* 1216, 2150–2155.

Vapnik, V.N. (1995) *The Nature of Statistical Learning Theory*. Springer, New York.

Wang, L., Lee, F.S.C., Wang, X. and He, Y. (2006) Feasibility study of quantifying and discriminating soybean oil adulteration in camellia oils by attenuated total reflectance MIR and fiber optic diffuse reflectance NIR. *Food Chemistry* 95, 529–536.

Wang, N., Sun, D. and Dong, H.P. (2009) Quick identification of shuanghuanglian oral liquid and yinhuang oral liquid by AOTF-near infrared spectroscopy principal component analysis. *Chinese Pharmaceutical Journal* 20, 2355–2357.

Wang, Y., Veltkamp, D.J. and Kowalski, B.R. (1991) Multivariate instrument standardization. *Analytical Chemistry* 63, 2750–2756.

Warren, M. and Pitts, W. (1943) A logical calculus of the ideas immanent in nervous activity. *Bulletin of Mathematical Biology* 5, 115–133.

Westad, F., Schmidt, A. and Kemit, M. (2008) Incorporating chemical band assignment in near infrared spectroscopy regression models. *Journal of the Near Infrared Spectroscopy* 16, 265–273.

Williams, P.C. (1987) Variables affecting near-infrared reflectance spectroscopic analysis. In: Williams, P. and Norris, K. (eds) *Near-infrared Technology in the Agricultural and Food Industries*. American Association of Cereal Chemists, St Paul, Minnesota, pp. 143–167.

Williams, P.C. (2001) Implementation of near infra technology. In: Williams, P.C. and Norris, K.H. (eds) *Near-infrared Technology in the Agricultural and Food Industries*. American Association of Cereal Chemists, St Paul, Minnesota, pp. 145–169.

Williams, P.C. and Norris, K.H. (eds) (2001) *Near-infrared Technology in the Agricultural and Food Industries*, Vol. 2. American Association of Cereal Chemists, St Paul, Minnesota.

Wilson, N.D., Watt, R.A. and Moffat, A.C. (2001) A near infrared method for the assay of cineole in Eucalpytus oil as an alternative to the official BP method. *Journal of Pharmacy and Pharmacology* 53, 95–102.

Wilson, N.D., Ivanova, M.S., Watt, R.A. and Moffat, A.C. (2002) The quantification of citral in lemongrass and lemon oils by near-infrared spectroscopy. *Journal of Pharmacy and Pharmacology* 54, 1257–1263.

Wold, S., Geladi, P., Esbensen, K. and Ohman, J. (1987) Multi-way principal components and PLS-analysis. *Journal of Chemometrics* 1, 41–56.

Wold, S., Sjostrom, M. and Eriksson, L. (2001) PLS-regression: a basic tool of chemometrics. *Chemometrics and Intelligent Laboratory Systems* 58, 109–130.

Wu, Y.J., Li, W., Xiang, B.R., Wu, Y.M., Su, P., Yuan, C.Q. and Yu, M.L. (2000) The application of near infrared reflectance spectroscopy technique in *Peucedaneae drude* plant taxonomy. *Computers and Applied Chemistry* 17, 111–112.

Xiang, L., Fan, G., Li, J., Kang, H., Yan, Y., Zheng, J. and Guo, D. (2002) The application of an artificial neural network in the identification of medicinal rhubarbs by near infrared spectroscopy. *Phytochemical Analysis* 13, 272–276.

Xing, J.S. and Zhang, X.B. (2010) Development of a near infrared method for rapid determination of ampicillin capsules. *Chinese Journal of Pharmaceutical Analysis* 28, 936.

Xu, D.L. and Ling, G.L. (2010) Construction of NIR models for determination of water in hard capsule of Chinese traditional medicine. *Chinese Journal of Pharmaceutical Analysis* 30, 2170–2172.

Yang, N. and Ren, G. (2008) Application of near infrared reflectance spectroscopy to the evaluation of rutin and D-chiro-inositol contents in tartary buckwheat. *Journal of Agricultural and Food Chemistry* 56, 761–764.

Yap, K.Y., Chan, S.Y., Weng, C.Y. and Sing Lim, C. (2005) Overview on the analytical tools for quality control of natural product based supplements: a case study of ginseng. *ASSAY and Drug Development Technologies* 3, 683–689.

Yuan, Y.F., Tao, Z.H., Liu, J.X., Tian, C.H., Wang, G.W. and Li, Y.Q. (2011) Identification of *Cortex phellodendri* by Fourier-transform infrared spectroscopy and principal component analysis. *Spectroscopy and Spectral Analysis* 31, 1258–1261.

Zalacain, A., Ordoudi, S.A., Diaz-Plaza, E.M., Carmona, M., Blazquez, I., Tsimidou, M.Z. and Alonso, G.L. (2005) Near infrared spectroscopy in saffron quality control: determination of chemical composition and geographical origin. *Journal of Agricultural and Food Chemistry* 53, 9337–9341.

Zhang, Q.H., Zhu, Z.W. and Li, J. (2011) Research progress on chemical composition and cultivation of TCM qinghao. *Journal of Traditional Chinese Medicine* 8, 10–12.

Zhang, Z., Wang, Y., Fan, G. and Harrington, P. De B. (2007) A comparative study of multilayer perceptron neural networks for the identification of rhubarb samples. *Phytochemical Analysis* 18, 109–114.

Zheng, H.J. and Han, Y. (2011) Establishment of near infrared qualitative model for tongxinluo capsules. *Chinese Pharmaceutical Affairs* 25, 373–374.

Zhou, J.L., Qi, L.W. and Li, P. (2008) Quality control of Chinese herbal medicines with chromatographic fingerprint. *Chinese Journal of Chromatography* 26, 153–159.

8 Antioxidant Assay

8.1 Introduction

Nutrition and health are interconnected, and regular consumption of fruit and vegetables is associated with many health benefits (Pellegrini *et al.*, 2003). They contain a wide variety of biologically active, non-nutritive compounds known as phytochemicals. A great number of plants worldwide have proved to have strong antioxidant activity and powerful scavenger activity against free radicals. Medicinal plants have also been studied extensively for their antioxidant activity. Some widely consumed beverages like tea, red wine and cocoa are well known for their high antioxidant activities. Antioxidants stop the reactions of radicals. As there are many possible radicals formed during oxidation, antioxidants must be effective against all radical pathways. Synthetic antioxidants are very effective in blocking specific radical reactions. However, they are far less effective than the antioxidant systems that occur naturally in plants, because plant-derived antioxidants block both the major and minor oxidation pathways.

Oxidation is a very complex process and oxidizable products contain different lipids. Plants use an array of molecules to suppress oxidation. The antioxidant activities of plants are attributed to a broad variety of substances like vitamins (ascorbic acid, tocopherols), polyphenolic compounds, carotenoids and other minor constituents (Liu *et al.*, 2008). Natural antioxidants such as vitamins A, C, E and β-carotene are abundant in vegetables (Wootton-Bearda *et al.*, 2010).

Flavonoids, flavones and polyphenolic compounds are non-nutritional antioxidants. Both vitamin C and E exhibit antioxidant activity under hydrophilic and lipophilic conditions. Vitamin E is located in cell membranes and is capable of reducing the free radicals in cell membranes. Vitamin E in vegetable oil is a complex of ten different antioxidant

 © Satyanshu Kumar 2016. *Analytical Techniques for Natural Product Research* (S. Kumar)

molecules. Plastoquinones, ubiquinones and tocotrienols are the three antioxidant systems related to the vitamin E family in plants. In addition, phaeophytins, sterols, sterol ferulates, carotenoids and phospholipids are other antioxidants that are present in oilseeds. β-Carotene is the major antioxidant in cell membranes and lipoproteins, although it shows a weaker antioxidant activity (Niki *et al.*, 1995). The beneficial health effects of fruit and vegetables are associated with different antioxidant compounds such as ascorbic acid (vitamin C), vitamin E, carotenoids, lycopenes, polyphenols and other phytochemicals present in fruit and vegetables (Prior and Cao, 2000). Most of the antioxidant capacity associated with fruit is owing to ascorbic acid and phenolic compounds. Only about 1% of the total antioxidant activity is due to lipophilic compounds (Wu *et al.*, 2004; Chavez-Santoscoy *et al.*, 2009). It has been established in research studies that these antioxidant components lower the risk of several diseases (Burton, 1989; Kahkonen *et al.*, 1999; Wargovich, 2000). Measuring antioxidant activity is a common practice to determine the potential inhibition or scavenging capacity of foods against reactive oxygen species (ROS).

Plants have been used for years as a source of traditional medicine to combat various diseases. Biological combustion involved in the respiration process produces ROS. Excess ROS in the body can lead to cumulative damage in proteins, lipids and DNA, resulting in oxidative stress. Oxidative stress is a result of an imbalance between the formation of ROS (oxidant) and antioxidants, in favour of oxidants (Sies, 1991). Interest in the search for new natural antioxidants has grown over the past years because ROS production and oxidative stress have been shown to be linked intricately with ageing-related diseases and longevity (Finkel and Holbrook, 2000). Oxygen-derived radicals represent the most important class among the free radicals formed in living organisms (Miller *et al.*, 1990).

Oxidative stress has been linked to various diseases (Halliwell, 1990). Oxidative damage to cells is one of several factors causing many diseases (Osawa *et al.*, 1995). Free radicals are ROS such as peroxyl radicals (ROO·), hydroxyl radicals (HO·), superoxide anions ($O_2\cdot$) and singlet oxygen ($^1O_2$). These are formed during cellular metabolism. The excess production of oxygen radical species such as hydrogen peroxide, superoxide anion radicals and hydroxyl radicals is thought to cause damage in cells (Wickens, 2001). More than 45 diseases including inflammation, cancer, cardiovascular and Alzheimer's disease and Parkinson's disease are known to be mediated by free radicals (Davies, 2000; Finkel and Holbrook, 2000; Tripathi and Upadhyay, 2002; Chavez-Santoscoy *et al.*, 2009). Although living cells possess a protective system of antioxidants that prevents the excessive formation and enables the deactivation of ROS, the prevention of oxidative stress by dietary supplementation of antioxidants has been the focus of research on antioxidant defence (Sasaki *et al.*, 2002; Tapiero *et al.*, 2002; Ren *et al.*, 2003). Antioxidants have been defined as any substance that, when present in low concentration compared to an oxidizable substance, significantly delays or prevents the oxidation of that substrate (Halliwell, 1990). Antioxidants are present in all plants and all parts of the plant (Co *et al.*, 2010). The presence of several antioxidant components such as flavonoids, flavonols, vitamins and tannins in plants make them an attractive new source of antioxidants to replace synthetic antioxidants. Butylated hydroxyl anisole (BHA), butylated hydroxyl toluene (BHT), propyl gallate and tert-butylhydroquinone (BHQ) are currently the most commonly used synthetic antioxidants (Amensour et al., 2010). However, concern over the safety of synthetic antioxidants has increased, and several countries have restricted their use because of their undesirable effects on human health (Chen et al., 1992; Kahl and Kappus, 1993). Synthetic antioxidants can generally cause problems of toxicity. Further, BHA and BHT have restricted use in food as they are reported to be carcinogenic, and some countries have not permitted the use of BHQ, the most potent synthetic food antioxidant, and other countries may ban it. Interest in the search for new natural antitoxins has grown over the past years, and demand is increasing for food antioxidants from natural sources (Shahidi, 1997; Suja et al., 2005). Antioxidants are of interest to the food industry because they prevent the formation of rancidity, deterioration and also nutritional losses (Fernandez-Lopez et al., 2005; Kuppusamy et al., 2002; Kratchanova et al., 2010). Antioxidants are of great interest in many commercial areas also. They can be added to prevent yellowing and also to minimize the evolution of odorous volatile gases in paper products of the paper and pulp industry (Fagerlund et al., 2003). Antioxidants are also valuable additives in polymers to prolong their shelf life (Malmstrom et al., 1998; Viscidi et al., 2004).

The concept of antioxidant capacity first originated from chemistry and was later adapted to biology, medicine, epidemiology and nutrition. It describes the ability of redox molecules in foods and biological systems to scavenge free radicals. This concept provides a broader picture of the antioxidants present in a biological sample as it considers the additive and synergistic effects of all antioxidants rather than the effect of single compounds and it may, therefore, be useful to study the potential health benefits of antioxidants on oxidative stress mediated diseases. The effectiveness of antioxidants is measured by monitoring the inhibition of oxidation of a suitable substrate, and the simplest tests of antioxidant activity involve the addition of an antioxidant to a model substrate. In general, antioxidant activity is associated with specific compounds or classes; they can undergo several deferment mechanisms and three classes of antioxidants have been defined (Scott, 1985). Preventive antioxidants avert the initiation or propagation of radical chain reactions and UV absorbers; metal chelators and peroxide decomposers belong to this group. Chain-breaking antioxidants react with an already formed reactive compound, turning it into a less reactive compound. Chain-breaking donor antioxidants typically reduce oxygen-centred radicals, and chain-breaking acceptor antioxidants typically oxidize alkyl radicals.

8.2 Antioxidant Capacity Measurement

Measurement of the antioxidant capacity of food products is a matter of growing interest because it may provide a variety of information such as resistance to oxidation, quantitative contribution of antioxidant substances or the antioxidant activity that they may present inside the organism when ingested (Huang *et al.*, 2005; Serrano *et al.*, 2007).

The various methods of measuring antioxidant activity differ in their reaction mechanism, the target substance on which they act, the conditions under which they are carried out and the way in which the results are expressed. These methods are used in different areas such as physiology, pharmacology, nutrition and food science, which renders it difficult to select the most appropriate method to provide the correct interpretation of results (Longhi *et al.*, 2011). The determination of antioxidant capacity should take into account the overall concentrations and compositions of diverse antioxidants, as the total antioxidant capacity is attributed to the combined activities of diverse antioxidants. The antioxidant assay is measured with a variety of tests in order to assess their ability to counteract the effects of various radicals. The requirement of a standard assay is very important in order to compare the results of different laboratories and to validate the conclusions. There is no official standardized method, and therefore it is recommended that each evaluation should be made with various oxidation conditions and different methods (Frankel and Meyer, 2000). A number of assays have been introduced to measure the total antioxidant activity of extracts, pure compounds and body fluids. Each method relates to the generation of a different radical acting through a variety of mechanisms and the measurement of a range of end points at a fixed time or over a range (Miller and Rice-Evans, 1994a). Two types of approach have been taken:

1. Inhibition assays in which the extent of scavenging by the hydrogen or electron donation of a preformed free radical is the marker of antioxidant activity.
2. Assays involving the presence of antioxidant systems during the generation of radicals.

The antioxidant capacity of foods depends on many factors, including the colloidal properties of the substrates, the condition and stages of oxidation and the localization of antioxidants in different phases (Frankel and Meyer, 2000). Also, the calculated antioxidant capacity of a sample depends on which technology and free radical generator or oxidant is used in the measurement. Therefore, the comparison of different analytical methods for determining total antioxidant capacity is a key factor in selecting a method and understanding the result. Numerous *in vitro* studies have been described to evaluate the total antioxidant capacity (TAC) of food products. As there is no official standardized method, therefore, it is recommended that each evaluation should be made with various oxidation conditions and different methods of measurement (Frankel and Meyer, 2000). Several assays, including 2,2-diphenyl-1-picrylhydrazyl (DPPH), 2,2-azobis(3-ethyl-benzothialzoline-6-sulfonic acid) (ABTS), ferric reducing antioxidant power (FRAP) and oxygen radical absorption capacity (ORAC) among others, are frequently used to estimate antioxidant capacities. Assays differ in terms of their principles and experimental conditions, and particular antioxidants have varying contributions to total antioxidant potential (Cao and Prior, 1998). The most commonly used antioxidant assay methods are DPPH and ABTS$^{+}\cdot$. Both methods are characterized by excellent reproducibility under certain assay conditions. They also show significant differences in their response to antioxidants. DPPH free radical (DPPH$\cdot$) does not require any special preparation, while ABTS radical cation must be generated. Also, ABTS$^{+}\cdot$ can be dissolved in aqueous and organic media in which the antioxidant activity can be measured due to the hydrophilic and lipophilic nature of the compounds in the sample. DPPH can only be dissolved in organic media, especially in methanol or ethanol, this being an important limitation when interpreting the role of hydrophilic antioxidants. Both radicals show similar bi-phasic kinetic reactions with many antioxidants.

In most assays, the same principle applies: a synthetic coloured radical or redox active compound is generated and the ability of the biological sample to scavenge the radical or to reduce the redox active compound is monitored by a spectrophotometer by applying an appropriate standard to quantify the antioxidant capacity, such as trolox equivalent capacity (TEAC) or vitamin C equivalent antioxidant capacity. Out of two approaches, one is based on electron transfer (ET) involving the reduction of a coloured oxidant; for example, ABTS, DPPH and FRAP assay. ET-based assays measure the capacity of an antioxidant in the reduction of an oxidant, which changes colour when reduced and the degree of colour change is correlated with the sample's antioxidant concentration. The other approach involves a hydrogen atom transfer (HAT), like ORAC assay, in which antioxidants and substrate compete for thermally generated peroxyl radicals. The majority of HAT-based assays are based on a competitive scheme in which antioxidants and substrates compete for thermally generated peroxy radicals through the decomposition of azo compounds.

8.3   *In Vitro* Antioxidant Assay

Antioxidant activity depends on the hydro-/lipophilicity of the antioxidant. The antioxidant activity of plant extract depends on the concentration of antioxidant phytochemicals and the solvent used for extraction, and also on its form of preparation. Furthermore, the complex compositions of the extracts could intensify certain interactions such as synergistic, additive or antagonistic effects between their components and/or medium (Koleva *et al.*, 2002). Antioxidant capacity as measured by different *in vitro* assays differs. Different assays have been introduced to measure antioxidant capacity in order to assess its ability to counteract the effects of various radicals.

8.3.1   2,2-Diphenyl-1-picrylhydrazyl (DPPH) assay

The relatively stable organic radical DPPH has been used widely in determination of the antioxidant activity of single compounds, as well as of different plant extracts (Katalinic *et al.*, 2006). DPPH assay uses a radical dissolved in organic media and is based on the reduction of the purple DPPH• to 1,1-diphenyl-2-picrylhydrazine with discoloration, and is therefore applicable to hydrophobic systems. DPPH•, a stable free radical with a characteristic absorption at 515 nm, is used to study the radical scavenging effects of plant extracts. The degree of discoloration indicates the scavenging potential of the sample antioxidant. As antioxidants donate protons to this radical, absorption decreases (Fig. 8.1). The decrease in absorption is taken as a measure of the extent of radical scavenging. The lower absorbance at 515 nm represents the higher DPPH scavenging activity. The percentage of DPPH scavenging activity is expressed by the following equation:

Percentage inhibition = [1 – (test sample absorbance/blank sample absorbance)] × 100.

The results are expressed in terms of EC_{50}, which corresponds to the amount of sample needed in terms of antioxidant concentration to decrease the initial DPPH• concentration by 50%. EC_{50} is calculated from a linear equation obtained by plotting antioxidant concentration and average percentage antioxidant activity (Table 8.1). The antiradical power (ARP) is another way to express the results, and ARP of extract is calculated as (Arbos, 2004; Suja *et al.*, 2005):

ARP = $(1/EC_{50})$.

High concentration of DPPH in the reaction mixture gives absorbance beyond the accuracy of the spectrophotometer measurement (Ayre, 1949; Sloane and William, 1977). As a result of widely different protocols being used by different research groups, the EC_{50} values for even the standard antioxidants like ascorbic acid and BHT vary. Light, oxygen and pH of the reaction mixture also affect the absorbance of DPPH (Ozcelik *et al.*, 2003). The accuracy range of spectrophotometric measurements falls within an absorbance of 0.221–0.698, which is equal to a transmittance of 20–60% (Ayre, 1949). DPPH concentration corresponding to this range is 25–70 µm

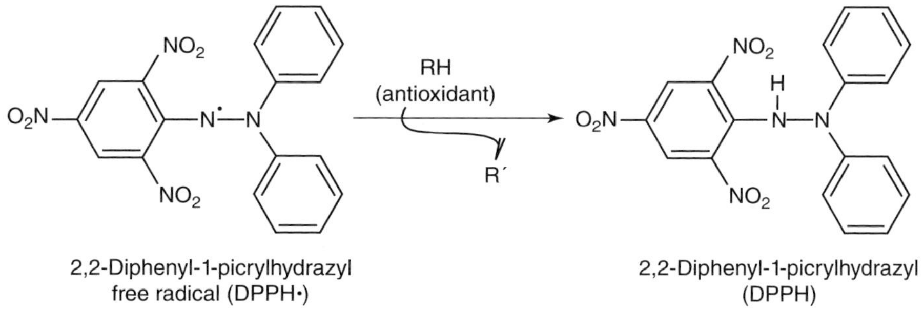

2,2-Diphenyl-1-picrylhydrazyl free radical (DPPH•)

2,2-Diphenyl-1-picrylhydrazyl (DPPH)

Fig. 8.1. Reaction between DPPH free radical to form DPPH.

Table 8.1. EC_{50} values of ascorbic acid, BHT and propyl gallate using DPPH antioxidant assay. (From Sharma and Bhatt, 2009.)

Standard	IC_{50} (µm)
Ascorbic acid	11.8,[a] 11.5[b]
BHT	60,[a] 9.7[b]
Propyl gallate	4.4,[a] 4.7[b]

[a]Methanol; [b]buffered methanol.

(Sharma and Bhatt, 2009). The absorbance of DPPH without any addition was stable over 30 min, and the suitable solvent for the DPPH assay was methanol or buffered methanol for the antioxidant assay of non-polar and polar compounds/extracts. DPPH radical scavenging activity is influenced by the polarity of the reaction medium, the chemical structure of the radical scavenger and the pH of the reaction mixture (Ozcelik *et al.*, 2003; Saito *et al.*, 2004; Shizuka and Kawata, 2005). The EC_{50} values of three standard antioxidants, ascorbic acid, BHT and propyl gallate, in methanol and buffered methanol was reported by Sharma and Bhatt (2009). The radical scavenging profiles of ascorbic acid and propyl gallate were similar in methanol and buffered methanol as solvents. However, unlike ascorbic acid and propyl gallate, the DPPH radical scavenging activity of BHT was markedly high in buffered methanol as compared to methanol alone. The difference in the IC_{50} values of BHT in methanol and buffered methanol could be due to one or more factors, as described above.

Longhi *et al.* (2011) reported the ARP of five standard antioxidants in the following order: BHA (2.11), vitamin C (3.70), α-tocopherol (3.85), quercetin (10.95) and gallic acid (12.72).

The DPPH scavenging activity of phenolics was correlated positively with the number of hydroxyl groups (Sroka and Cisowski, 2003). This observation explains the relative EC_{50} values of BHT and propyl gallate. Propyl gallate with three hydroxyl groups has lower IC_{50} values as compared to BHT with one hydroxyl group. The radical scavenging reaction of ascorbic acid with DPPH was instantaneous. On the other hand, the radical scavenging reaction of BHT with DPPH was slow and absorbance continued to decrease for a period of 90 min of observation. The reaction of DPPH with propyl gallate was also quite fast, but slower as compared to that with ascorbic acid. For the sake of uniformity, a time interval of 30 min was taken for ascorbic acid, BHT and propyl gallate radical scavenging capacity measurements. However, it is important to carry out a time course of radical scavenging activity while using DPPH radicals for the assay of antioxidant assay.

The DPPH method is very rapid, simple, sensitive and reproducible, and does not require special instrumentation. Very mild experimental conditions are required for this assay. DPPH assay is not discriminative with respect to radical species, but gives a general idea of the radical quenching ability.

8.3.2 2-2′-Azinobis-(3-ethylbenzothiazoline-6-sulfonic acid)

2,2′-Azinobis-(3-ethylbenzothiazoline-6-sulfonic acid) (ABTS) has high water solubility and chemical stability. It is a peroxidase substrate and generates a metastable radical with a characteristic absorption spectrum and high molar absorptivity at 414 nm when oxidized in the presence of hydrogen peroxide (Arnao *et al.*, 1996). However, there are secondary absorption maxima in the wavelength regions of 645, 734 and 815 nm. ABTS assay is based on the generation of a blue/green ABTS·+ that can be reduced by antioxidants in the reaction medium on a timescale dependent on the antioxidant activity. This ABTS radical cation decolonization assay is applicable to both lipophilic and hydrophilic antioxidants, including flavonoids, hydroxycinnamates, carotenoids and plasma antioxidants. Generation of the ABTS (2,2′-azinobis(3-ethylbenzothiazoline-6-sulfonic acid) radical cation forms the basis of this spectrophotometric antioxidant assay. The preformed radical mono cation of ABTS·+ is generated by the oxidation of ABTS with potassium persulfate and is reduced in the presence of hydrogen donating antioxidants (Fig. 8.2). The addition of antioxidants to the preformed radical cation reduces it to ABTS to an extent and on a timescale depending on the antioxidant activity, the concentration of the antioxidants and the duration of the reaction. The extent of decolonization as the percentage inhibition of the ABTS·+ radical cation is determined as a function of concentration and

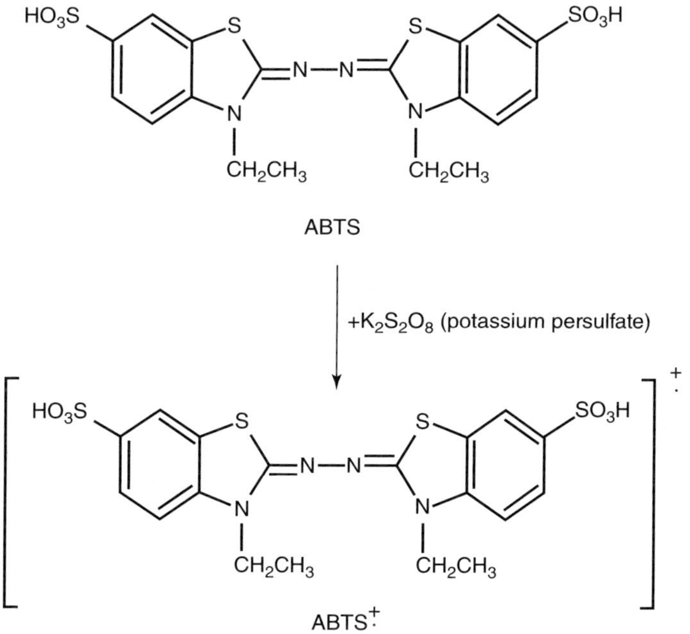

Fig. 8.2. Reaction for formation of ABTS·+ radical.

time, and is calculated relative to the reactivity of trolox as standard and under the same conditions. ABTS is dissolved in water to a 7-mM concentration. ABTS radical cation (ABTS·$^+$) is produced by reacting ABTS solution with 2.45 mM potassium persulfate and allowing the mixture to stand in the dark at room temperature for 12–16 h before use.

For the study of phenolic compounds and food extracts, the ABTS·$^+$ solution is diluted with ethanol, and for plasma antioxidants with 5 mM phosphate buffered saline (PBS) pH 7.4, to an absorbance of 0.70 ± 0.02 at 734 nm and equilibrated at 30°C. After the addition of 1.0 ml of diluted ABTS·$^+$ solution (A_{734} = 0.70 ± 0.02) to 10 µl of antioxidant compounds or trolox standards (final concentration = 0–15 µM) in ethanol or PBS, the absorbance reading was recorded at 30°C exactly 1 min after initial mixing and up to 6 min. Appropriate solvent blanks are run in each assay. All determinations are carried out at least three times and in triplicate. The reduction of the absorbance is calculated according to the following equation:

$$\text{Percentage inhibition} = (\text{Abs}_{t=0} - \text{Abs}_{t=t})/\text{Abs}_{t=0} \times 100,$$

where $\text{Abs}_{t=0}$ = absorbance at 0 min; $\text{Abs}_{t=t}$ = absorbance after t min.

The percentage inhibition of absorbance at 734 nm is calculated and plotted as a function of concentration of antioxidants and trolox for the standard and the reference data (Re *et al.*, 1999). The reduction of the absorbance is plotted against the amount of sample to draw a regression line. The ratio between the sample and trolox's slope of regression line is calculated and expressed as the TEAC. Results are expressed as TEAC (micromole trolox equivalents per milligram of dry extract). Higher TEAC values correspond to higher antioxidant activity.

Both DPPH and ABTS are convenient in their application and large number of samples can be screened in a short time; nevertheless, they are limited as they use non-physiological radicals. The results of DPPH and ABTS assays differ for plant extracts as the stiochiometry of reactions between the antioxidant compounds in the extract and ABTS·$^+$ and DPPH· differ. Also, the electron reduction potential of ABTS·$^+$ and DPPH· is different for different compounds (Amensour *et al.*, 2010).

Relative antioxidant capacity (RACI) was defined as an integrated approach to compare the antioxidant capacity of different foods or food components measured with two or more chemical assays (Sun and Tanumihardjo, 2007; Sun *et al.*, 2009).

8.3.3 Phosphomolybdenum method

The phosphomolybdenum method is a quantitative method and total antioxidant activity is expressed as the number of equivalents of ascorbic acid. Spectrophotometric measurement of antioxidant capacity by the phosphomolybdenum method is based on the reduction of molybdenum (VI) to molybdenum (V) by the analyte sample (Prieto *et al.*, 1999). Sample solution is mixed with a reagent solution comprising sulfuric acid (0.6 M), sodium phosphate (28 mM) and ammonium molybdate (4 mM) and incubated in a boiling water bath for 90 min. Absorbance of green coloured phosphate/molybdenum (V) compounds with absorption maxima at 695 nm is measured. For samples of unknown composition, water-soluble antioxidant capacity is expressed as the equivalent of ascorbic acid, µmole g^{-1} of extract (Gupta and Prakash, 2009).

8.3.4 Reducing power assay

Reducing power has been used as one of the antioxidant capability indicators of medicinal herbs (Duh and Yen, 1997). Reducing properties are generally associated with the presence of reductones, which are believed to break radical chains by the donation of a hydrogen atom, indicating that antioxidative properties are concomitant with the development of reducing power (Gordon, 1990). The reducing power may accord with the overall antioxidant activity. Tanaka *et al.* (1988) noted that the antioxidative effect increased exponentially as a function of the development of the reducing power, indicating that the antioxidative properties were concomitant with the development of the reducing power.

In this assay, the Fe^{3+}/ferricyanide complex is reduced to the ferrous form (Fe^{2+}) by antioxidants. The ferrous ion formed is monitored by measuring the formation of Perl's Prussian blue at 700 nm (Oyaizu, 1986). The higher the absorbance, the stronger the reducing power (Guo *et al.*, 2001). Reducing power is increased by increasing the sample concentration.

Oxidant

Fe(S-CN)$_2$ $\longrightarrow$ Fe(S-CN)$_3$

Ferrous thiocyanate   Fe^{3+} thiocyanate

8.3.5 Ferric reducing antioxidant power

Ferric reducing antioxidant power (FRAP), also, known as ferrous ion chelating ability assay, is different from other assays as no free radicals are involved, but the reduction of ferric iron (Fe^{3+}) to ferrous iron (Fe^{2+}) is monitored. Ferrous ion, commonly found in food systems, is well known as an effective pro-oxidant due to its high reactivity. Ferrous ion participates in the direct or indirect initiation of lipid oxidation (Wettasinghe and Shahidi, 2002).

FRAP assay is carried out to determine the capability of a substance to bind with the oxidation catalytic ferrous ion. The FRAP method is based on the reduction of a colourless ferroin analogue, Fe^{3+} complex of tripyridyltriazine Fe (TPTZ)$^{3+}$, to the intensely blue coloured ferrous (Fe^{2+}) complex, Fe (TPTZ)$^{2+}$, by antioxidants in acidic media (at low pH). A complex of Fe^{2+}/ferrozine has a strong absorbance at 562 nm.

Ferrozine can form complexes with Fe^{2+} quantitatively. The complex formation is disrupted in the presence of other chelating agents, with the result that the colour of the complex decreases.

A measure of the rate of colour reduction allows estimation of the chelating activity of the coexisting chelator (Yamaguchi, 1980). The higher the ferrous ion chelating ability of the test sample gives the lower absorbance. The percentage inhibition of the formation of the Fe^{2+}/ferrozine complex is calculated using the formula:

Scavenging effect (%) = $[(A_{control} - A_{sample})/A_{control}] \times 100$,

where $A_{control}$ = absorbance of the Fe^{2+}/ferrozine complex and A_{sample} = absorbance of the test compound.

The percentage inhibition of absorbance is plotted as a function of concentration of ethylenediamine tetra acetic acid (EDTA) for the standard reference data. The results are expressed in terms of EDTA equivalents (micromole EDTA equivalents per gram of dry extract).

The FRAP assay is quick and simple to perform to measure the antioxidant capacity not only of pure compounds but also of fruit, wines and animal tissues (Wojdyło et al., 2007). The reference antioxidant for this assay must be water soluble, such ascorbic acid, uric acid or trolox. This method is used mostly in conjugation with other assays. However, the reducing capacity does not necessarily reflect the antioxidant activity (Katalinic et al., 2006; Wong et al., 2006).

8.3.6   Oxygen radical absorbance capacity

Oxygen radical absorbance capacity (ORAC) has been found to be the most relevant method for biological samples. ORAC measures the absorbance capacity of peroxyl radicals. The ORAC method is the only one so far that combines the total inhibition time and the percentage of free radical damage by the antioxidant into a single quantity (Zulueta et al., 2009). The method is based on the detection of chemical change in a fluorescent molecule caused by a free radical attack. It measures the antioxidant scavenging activity against peroxyl radicals generated by thermal decomposition of 2,2-azobis(2-amidinopropane) dihydrochloride (AAPH) (Fig. 8.3). AAPH is used as a peroxyl radical generator, fluorescein (FL), as a fluorescent probe and trolox as a standard. The loss of fluorescence of FL is an indication of the extent of damage caused from its reaction with the peroxyl radicals. The ORAC assay is carried out on a multi-label counter with fluorescence filters. Initially developed by Cao et al. (1993), the method consists of measuring the decrease in the fluorescence of a protein as a result of the loss of its conformation when it suffers oxidative damage caused by a source of peroxyl radicals (ROO·). The method measures the ability of the antioxidants in the sample to protect protein from oxidative damage. The protective effect of an antioxidant is measured by assessing the area under the fluorescence decay curve, and antioxidant activity is

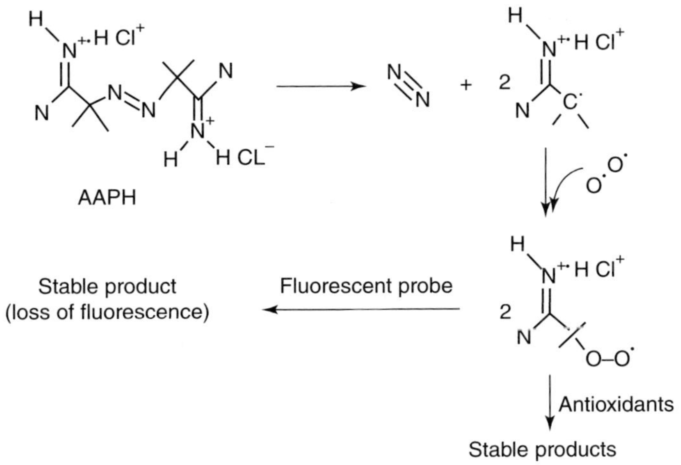

Fig. 8.3. Reaction of the AAPH radical during the ORAC assay.

expressed in micromole trolox equivalents per gram of dry weight of food or biological samples (Kratchanova *et al.*, 2010). ORAC is extremely sensitive. The samples must be diluted appropriately before analysis to avoid interference.

8.3.7 β-Carotene–linoleate method

Real food systems consist of multiple phases in which lipid and water co-exist with some emulsifier; therefore, an antioxidant assay using a heterogeneous system such as oil in water emulsion is also required. Linoleic acid and linoleic acid emulsion systems represent homogeneous and heterogeneous systems (Osawa and Namiki, 1981). A linoleic acid system can be correlated with a homogeneous systems or bulk oil phase system. Also, a linoleic acid emulsion system can be simulated with a biological lipid system, or with food or fat emulsion. In β-carotene–linoleic acid bleaching and linoleic acid emulsion system–thiocyanate methods, inhibition of peroxidation is taken as the index of activity. In a β-carotene–linoleate model of an antioxidant assay system, the mechanism of bleaching of β-carotene is a free radical-mediated phenomenon resulting from the hydroperoxides formed from linoleic acid. β-carotene in this model system undergoes rapid discoloration in the absence of an antioxidant because of the coupled oxidation of β-carotene and linoleic acid generating free radicals. The linoleic acid free radical formed on the abstraction of a hydrogen atom from one of its diallylic methylene groups attacks the highly unsaturated β-carotene molecules and as they lose their double bond, β-carotene loses its chromophore and characteristic orange colour, which can be monitored spectrophotometrically. The presence of antioxidants can hinder the extent of β-carotene bleaching by neutralizing the linoleate free radical and other free radicals formed in the system (Jayaprakash *et al.*, 2001). In an antioxidant assay system using a linoleic acid–thiocyanate system, linoleic hydroperoxides are generated because of the oxidation of the linoleic acid, which further decomposes to many secondary oxidation products (Hua-Ming *et al.*, 1996). The oxidized products react with ferrous sulfate to form ferric sulphate, then to ferric thiocyanate of a blood-red colour. In the presence of antioxidants, the oxidation of linoleic acid is slowed; therefore, colour development due to the formation of thiocyanate will be slow.

The results are expressed in percentage basis of preventing bleaching of β-carotene. The antioxidant activity of the extracts is calculated in terms of β-carotene bleaching using the following formula (Abdille *et al.*, 2005):

$$\text{Antioxidant activity} = 100\,[1 - (A_0 - A_t)/(A_0^0 - A_t^0)],$$

where A_0 and A_0^0 are the absorbance values measured at zero time of the incubation for the test sample and control, respectively.

A_t and A_t^0 are the absorbance values measured at zero time of the incubation for the test sample and control, respectively, after incubation for 120 min. The method is sensitive due to the strong absorption of β-carotene, but is slower than the DPPH method. This assay has poor reproducibility because of variations in the β-carotene bleaching reaction. It is not specific as it is subject to interference from oxidizing and reducing agents in the extracts, and also, linoleic acid is not representative of typical food lipids (Spigno and De Faveri, 2007).

8.3.8   Thiobarbituric reactive substance (TBARS) assay

Thiobarbituric reactive substance (TBARS) assay is one of most popular assays for studies related to lipid peroxidation and it is used widely to evaluate the antioxidant activities of various natural products. Thiobarbituric acid reacts with many different carbonyl compounds formed from lipid peroxidation. Their TBA adducts absorb the same UV wavelength absorbed by the malonyl-thiobarbituric acid adduct.

Various concentrations of testing samples are added to an aqueous solution containing tris buffer (pH 7.4), potassium chloride (1 M), sodium dodecyl sulfate (SDS) (1%) and cod liver (this can be any kind of lipid such as linoleic acid, arachidonic acid or ω-3 fatty acids), ferrous chloride $FeCl_2$ and hydrogen peroxide (H_2O_2, 0.5 μM) in a non-transparent vial. The sample is incubated at 37°c for 18 h, with shaking. After the incubation, oxidation is terminated by adding BHT (4% in ethanol) solution and TBA reagent (0.67% TBA, trichloroacetic acid (TCA) 1%, SDS, 5N HCl) is added to the sample. The sample is heated at 80°C for 1 h and then cooled in an ice bath for 10 min. A blank is prepared following the same procedure without a test sample. In the thiobarbituric acid assay, malonaldehyde TBA-MA adduct product formed is measured using a spectrophotometer at 532 nm. A known antioxidant such as BHT, vitamin A or vitamin C is used as a positive control in the assay (Moon and Shibamoto, 2009).

8.3.9 Superoxide dismutase (SOD) assay

The superoxide anion scavenging activity of plant extracts is determined by the WST (2-(4-iodophenyl)-3-(4-nitrophenyl)-5-(2,4-disulfophenyl)-2H-tetrazolium monosodium salt) reduction method using the SOD assay kit-WST. Here, $O_2^{\cdot-}$ reduces WST to produce the yellow formazan measured spectrophotometrically. Antioxidants inhibit the formation of yellow WST (Dudonné et al., 2009).

8.3.10 Electron spin resonance (ESR) spectroscopy

Electron spin resonance (ESR) spectroscopy determines the presence of unpaired electrons of oxygen and is commonly used for free radical evaluation. Superoxide anion ($O_2^{\cdot-}$) scavenging capacities are measured by ESR assay (Tabart et al., 2009). It has been applied to some foods to measure free radical production and to establish their antioxidant capacities (Noda et al., 1997).

8.4 Antioxidant Capacity and Cooking and Storage Processes

Culinary herbs are a potential source of antioxidants in the diet. Their antioxidant capacity is believed to be primarily responsible for the protective effect against cardiovascular disease and cancer. Many culinary herbs are cooked or undergo some other form of processing before they are consumed as part of a meal and such factors may affect their significance as a source of dietary antioxidants. Chohan et al. (2008) reported the impact of cooking (simmering, microwaving, stewing, stir-frying and grilling) and storage (vinegar maceration, cold maceration and freezing) on the antioxidant capacity of some common culinary herbs such as cinnamon, cloves, sweet fennel seeds, lavender, parsley, red roses, rosemary, sage, thyme and ginger. Extracts were prepared pre or post cooking or storage and their antioxidant capacities determined. It was found that simmering, soup making and the stewing process increased their antioxidant capacity significantly, while grilling and stir-frying decreased it. Both freezing herbs at −20°C and cold maceration had preservative effects on antioxidant capacity. Choi et al. (2006) also reported that under the influence of heat, the antioxidant capacity of shiitake mushrooms was enhanced due to the triggering of a breakdown of the insoluble sections of the cell wall. This breakdown increased the pool of bioavailable polyphenolic compounds in the diet. The antioxidant capacity of grape seed extracts was also increased through the liberation of phenolic compounds by heat (Kim *et al.*, 2006). Choi *et al.* (2006) also suggested that Maillard reaction products, which increase the oxygen-scavenging activities in various products, might also be responsible for the increase.

Different assays have been introduced to measure the antioxidant capacity of foods and biological samples The antioxidant capacity of foods may differ depending on environmental factors such as the growing season, geographical origin and agricultural practices, because these factors can affect the accumulation of antioxidant components in plant material significantly (Chun *et al.*, 2005). Also, the measured antioxidant capacity of a sample depends on which technology and which free radical generator or oxidant is used in the measurement. No single method is adequate for evaluating the antioxidant capacity of foods, as the different methods can yield widely diverging results (Frankel and Meyer, 2000). Various methods based on different mechanisms must be used. Determination of the antioxidant capacity of food should take into account the overall concentrations and compositions of diverse antioxidants because the total antioxidant capacity is due to the combined activities of diverse antioxidants (Tabart *et al.*, 2009).

8.5   Structure and Antioxidant Activity Relationship

Distinct association between bioactive compounds and antioxidant activity might be related to the presence of various active compounds in the plant. The synergistic effects of different compounds, the experimental conditions and the mechanism of the antioxidant reactions method used may affect this association. Moreover, most methods have their own limitations in the determination of antioxidant activity (Jayaprakash and Patil, 2007).

Phenolic compounds are the most prevalent antioxidant phytochemicals in the plant kingdom. There are about 5000 known plant phenolics, and many of them exhibit significant antioxidant activity, both singlet oxygen-quenching activity and radical scavenging activity (Guo *et al.*, 1999; Robards *et al.*, 1999). However, there is a wide degree of variation between different phenolic compounds in their effectiveness as antioxidants, which is determined by several structural features (Ou *et al.*, 2002). Also, there are a number of different mechanisms by which phenolics may act as antioxidants: via free radical scavenging, hydrogen donation, singlet oxygen quenching, metal ion chelation or as a substrate for attack by superoxide (Hamilton *et al.*, 1997; Rice-Evans *et al.*, 1997a; Robak and Gryglewski, 1998). The antioxidant activity of phenolic compounds is due mainly to their redox properties, which allow them to act as reducing agents, hydrogen donors, singlet oxygen quenchers, heavy metal chelators and radical quenchers (Kaur and Kapoor, 2002).

The numbers and position of the hydrogen-donating hydroxyl groups on the aromatic ring of the phenolic molecules control the free radical and antioxidant activity of the phenolics. This is also affected by other factors such as the glycosylation of aglycones and other H-donating groups (-NH, -SH), etc. For phenolic acids (hydroxyl benzoic acid, hydroxyphenyl acetic and hydroxycinnamic acids) and their ester derivatives, it is known that antioxidant activity depends on the number of hydroxyl groups in the molecule that are affected by steric hindrance from their carboxylate group (Rice-Evans *et al.*, 1997a). The antioxidant activity of phenolic acids increases with additional hydroxyl groups. The closeness of the carboxylate group and hydroxyl groups on the phenolic ring in hydroxybenzoic acids affects their donor proton ability negatively. As a result, higher antioxidant activities are usually observed on hydroxycinnamic acids such as coumaric, caffeic and ferulic acid as compared to their hydroxybenzoic acid counterparts (Fig. 8.4). Further, ortho substitution with electron donating alkyl or methoxy groups increases the stability of the aryloxyl radical, and hence its antioxidant potential (Rice-Evans *et al.*, 1996a). The alkoxyl radical scavenging ability in a lipid peroxidation system increased in the following order: salicylic < vanillic < chlorogenic < caffeic < gallic acids (Milic *et al.*, 1998).

The antioxidant activities of flavonoids (flavonols, isoflavones, etc.) that have a diphenylpropane skeleton depend on the structure, degree of hydoxylation and substitution pattern of hydroxyl groups. Antioxidant potency is related to structure in terms of electron delocalization of the aromatic nucleus. Hydroxylation of the B-ring is the major requirement for consideration of activity (Herrmann, 1976; Miller, 1996b). Hydroxyl radical scavenging activity increases with the number of hydroxyl groups substituted on the B-ring, especially at C-3' (Ratty and Das, 1988).

A single hydroxyl substituent generates little or no antioxidant activity. Flavonones such as naringenin and hesperitin with only one hydroxyl group on the B-ring have negligible antioxidant activity. But all flavonoids with 3',4'-dihydroxy substitution possess antioxidant activity (Dziedzic and Hudson, 1983). Quercetin and cyanidin with 3',4'-dihydroxy substitution in the B-ring and conjugation between the A- and B-rings have antioxidant potential four times that of trolox. The essential requirement

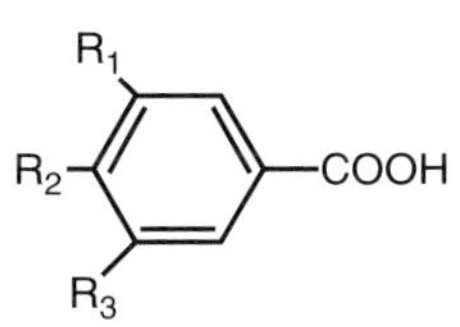

Benzoic acids
Gallic acid $R_1=R_2=R_3=OH$
Protocatechuic acid $R_1=H$, $R_2=R_3=OH$
Vanillic acid $R_1=H$, $R_2=OH$, $R_3=OCH_3$
Syringic acid $R_2=OH$, $R_1=R_3=OCH_3$

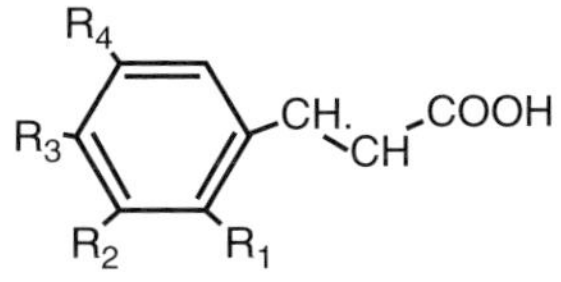

Cinnamic acids
Ferulic acid $R_1=R_2=H$, $R_3=OH$, $R_4=OCH_3$
p-Coumaric acid $R_1=R_2=R_4=H$, $R_3=OH$
o-Coumaric acid $R_2=R_3=R_4=H$, $R_1=OH$
Caffeic acid $R_1=R_2=H$, $R_3=R_4=OH$
Sinapic acid $R_1=H$, $R_3=OH$, $R_2=R_4=OCH_3$
Cinnamic acid esters

Fig. 8.4. Structure of phenolic compounds.

for effective radical scavenging is the 3′,4′-orthodihydroxy configuration in B-ring and the 4-carbonyl group in C-ring. The presence of 3-OH groups or 3- and 5-OH groups giving a catechol-like structure in C-ring is also beneficial for the antioxidant activity of flavonoids. The presence of the C-2–C-3 double bond configured with a 4-keto arrangement is known to be responsible for electron delocalization from B-ring, and it increases the radical scavenging activity. In the absence of the *O*-dihydroxy structure in B-ring, a catechol structure in A-ring can compensate for flavonoid antioxidant activity. The presence of glycosylations on the molecule may decrease its antioxidant activity.

The relationship between the chemical structure of flavonoids and their radical scavenging activities was analysed by Bors *et al.* (1990). Quercetin has a catechol structure in B-ring as well as a 2,3-double bond in conjuction with a 4-carbonyl group in C-ring, allowing for delocalization of the phenoxyl radical electron to the flavonoid nucleus. The combined presence of a 3-hydroxy group with a 2,3-double bond additionally increases the resonance stabilization for electron delocalization; hence, it has a higher antioxidant value. Quercetin and luteolin have an identical number of hydroxyl groups with 3′,4′- and 5,7-dihydroxyl groups in B- and A-rings, respectively. Flavonols (quercetin, myricetin, kaempferol and isorhamnetin) have a hydroxyl group in position 3. The 3,4-position of dihydroxylation of the phenolic ring in caffeic acid showed increased antioxidant activity as compared to *p*-coumaric acid (Kim and Lee, 2004). Caffeic acid is expected to have higher antioxidant activity because of additional conjugation in the propenoic side chain, which might facilitate electron delocalization by resonance between the aromatic ring and the propenoic group.

A linear correlation between the content of total phenolic compounds and their antioxidant capacity has been demonstrated (Cai *et al.*, 2004; Katsube *et al.*, 2004; Djeridane *et al.*, 2006; Katalinic *et al.*, 2006).

8.6   *In Vivo* Antioxidant Assay

In *in vivo* antioxidant assay, samples are usually administered to the testing animal (mice, rats, etc.) at a fixed dosage regimen by the respective method. After a specified time, the animals are sacrificed and blood or tissues are used for the assay (Alam *et al.*, 2013). A brief list of *in vivo* assays is given in Table 8.2.

8.7   Conclusion

Foods are no longer evaluated just in terms of nutrient composition but are also considered as promoters of good health. Antioxidants that have been shown to neutralize free radicals may be of central importance in the prevention of a number of diseases. Antioxidants are of interest to

Table 8.2. *In vivo* antioxidant assays.

Serial No.	Assay	Reference
1	Ferric reducing ability of plasma	Benzie and Strain, 1996
2	Reduced glutathione (GSH) assay	Elllman, 1959
3	Glutathione peroxidase (GSHPx) assay	Wood, 1970
4	Glutathione-S-transferase (GSt) assay	Jocelyn, 1972
5	Superoxide dismutase (SOD) assay	McCord and Fridovich, 1969
6	Catalase assay	Aebi, 1984
7	γ-Glutamyl transpeptidase (GGT) assay	Singhal *et al.*, 1982
8	Glutathione reductase assay	Kakkar *et al.*, 1984
9	l ipid peroxidation assay	Okhawa, 1979
10	LDL assay	El-Saadani *et al.*, 1989

researchers in food science as well as to health professionals. As the role of antioxidants in the diet and their impact on human health is coming under scrutiny, there has been a convergence of interests among researchers from these areas. A simple universal method by which total antioxidant capacity can be measured accurately and quantitatively does not exist. Different methods have been used to measure antioxidant capacity. Also, a wide range of spectrophotometric assays have been adapted to measure antioxidant capacity. Although DPPH and ABTS assays are convenient in their application, and thus are the most popular, they are, nevertheless, limited as they use non-physiological radicals. Further, ABTS and DPPH assays do not assess all of the antioxidant activities. A new concept called RACI has been developed by Sun and Tanumihardjo (2007) to integrate the results from multiple methods.

In vivo assays are suggested to be better in terms of understanding the interactions of phytochemicals, as the *in vitro* antioxidant activity does not necessarily predict the biological effectiveness of phytochemicals and extracts (Danesi *et al.*, 2008). Owing to the complexity of the oxidation–antioxidation processes, no single assay method is capable of providing a comprehensive picture of the antioxidant profile. The chemical complexity of plant extracts hampers the explanation and interpretation of their antioxidant activity. Various methods based on different mechanisms must be used in parallel to evaluate the antioxidant capacity, as different methods can give widely divergent results. Further, to obtain useful data, antioxidants should be studied in an environment similar to the real-life situation.

References

Abdille, M.H., Singh, R.P., Jayaprakash, G.K. and Jena, B.S. (2005) Antioxidant activity of the extracts from *Dillenia indica* fruit extracts. *Food Chemistry* 90, 891–896.
Aebi, H. (1984) Catalase. *Methods in Enzymology* 105, 121–126.

Alam, M.N., Bristi, N.J. and Rafiquzzaman, M. (2013) Review on *in vivo* and *in vitro* methods evaluation of antioxidant activity. *Saudi Pharmaceutical Journal* 21, 143–152.

Amensour, M., Sendra, E., Perez-Alvarez, J.A., Skali-Senhaji, N., Abrini, J., *et al.* (2010) Antioxidant activity and chemical content of methanol and ethanol extracts from leaves of Rockrose (*Cistus ladaniferus*). *Plant Foods for Human Nutrition* 65, 170–178.

Arbos, K.A. (2004) Estudo do potencial antioxidante de vegetais da família Cruciferae de diferentes cultivos. Curitiba, 86 pp. [Dissertação de Mestrado em Ciências Farmacêuticas. Setor de Ciências da Saúde. Universidade Federal do Paraná].

Arnao, M.B., Cano, A., Hernandez-Ruiz, J., Garcia-Canovas, F. and Acosta, M. (1996) Inhibition by L-ascorbic acid and other antioxidants of the 2-2′-azino-bis (3-ethylbenzthiazoline-6-sulfonic acid) oxidation catalyzed by peroxidase – a new approach for determining total antioxidants status of foods. *Analytical Chemistry* 236, 255–261.

Ayres, G.H. (1949) Evaluation of accuracy in photometric analysis. *Analytical Chemistry* 21, 1725–1729.

Benzie, I.F.F. and Strain, J.J. (1996) Ferric reducing antioxidant power assay: direct measure of total antioxidant activity of biological fluids and modified version for simultaneous measurement of total antioxidant power and ascorbic acid concentration. *Methods in Enzymology* 299, 15–27.

Bors, W., Heller, W., Michel, C. and Saran, M. (1990) Flavonoids as antioxidants: determination of radical-scavenging efficiencies. *Methods in Enzymology* 186, 343–355.

Burton, G.W. (1989) Antioxidant action of carotenoids. *Journal of Nutrition* 119, 109–111.

Cai, Y., Luo, Q., Sun, M. and Corke, H. (2004). Antioxidant activity and phenolic compounds of 112 traditional Chinese medicinal plants associated with anticancer. *Life Science* 74, 2157–2184.

Cao, G. and Prior, R.L. (1998) Comparison of different analytical methods for assaying total antioxidant capacity of human serum. *Clinical Chemistry* 44, 1309–1315.

Cao, G., Alessio, H.M. and Cutler, R.G. (1993) Oxygen radical absorbance capacity assay for antioxidants. *Free Radical and Biology Medicine* 14, 303–311.

Chavez-Santoscoy, R.A., Gutierrez-Uribe, J.A. and Serna-Saldivar, S.O. (2009) Phenolic composition, antioxidant capacity and *in vitro* cancer cell cytotoxicity of nine prickly pear (*Opunita* spp.) juices. *Plant Foods for Human Nutrition* 64, 146–152.

Chen, Q., Shi, H. and Ho, C.T. (1992) Effects of rosemary extracts and major constituents on lipid oxidation and soybean lipoxygenase activity. *Journal of American Oil Chemists Society* 103, 6472–6477.

Chohan, M., Forster-Wilkins, G. and Opara, E.I. (2008) Determination of the antioxidant capacity of culinary herbs subjected to various cooking and storage processes using ABTS[+] radical cation assay. *Plants Food for Human Nutrition* 63, 47–52.

Choi, Y., Lee, S.M., Chun, J., Lee, H.B. and Lee, J. (2006) Influence of heat treatment on the antioxidant activities and polyphenolic compounds of shitake (*Lentinus edodes*) mushroom. *Food Chemistry* 99, 381–387.

Chun, O.K., Kim, D.O., Smith, N., Schroeder, D., Han, J.T., *et al.* (2005) Daily consumption of phenolics and total antioxidant capacity from fruits and vegetables in the American diet. *Journal of the Science of Food and Agriculture* 85, 1715–1724.

Co, M., Fagerlund, A., Engman, L., Sunnerheim, K., Sjoberg, P.J.R., *et al.* (2010) Extraction of antioxidants from spruce (*Picea abies*) bark using eco-friendly solvents. *Phytochemical Analysis* 23, 1–11.

Danesi, F., Elementi, S., Neri, R., Maranesi, M., D'Antuono, L.F., *et al.* (2008) Effect of cultivar on the protection of cardiomyocytes from oxidative stress by essential oils and aqueous extracts of basil (*Ocimum basilicum* L.). *Journal of Agricultural and Food Chemistry* 52, 48–54.

Davies, K.J. (2000) Oxidative stress, antioxidant defenses and damage removal, repair and replacement systems. *IUBMM Life* 50, 279–289.

Djeridane, A., Yosuf, M., Nadfemi, B., Boutassouna, D., Stocker, P., *et al.* (2006) Antioxidant activity of some Algerian medicinal plants extracts containing phenolic compounds. *Food Chemistry* 97, 654–660.

Dudonné, S., Vitrac, X., Coutière, P., Woillez, M. and Mérillon, J.M. (2009). Comparative study of antioxidant properties and total phenolic content of 30 plant extracts of industrial interest using DPPH, ABTS, FRAP, SOD and ORAC assays. *Journal of Agricultural and Food Chemistry* 57, 1768–1774.

Duh, P. and Yen, G. (1997) Antioxidative activity of three herbal water extracts. *Food Chemistry* 60, 639–645.

Dziedzic, S.Z. and Hudson, B.J.F. (1983) Polyhydroxy chalcones and flavanones as antioxidants for edible oils. *Food Chemistry* 12, 205–212.

Ellman, G.L. (1959) Tissue sulfhydryl groups. *Archives of Biochemistry and Biophysics* 82, 70–77.

El-Saadani, M., Esterbauer, H., El-Sayeed, M., Goher, M., Nassar, A.Y., *et al.* (1989) Spectrophotometric assay for lipid peroxides in serum lipoproteins using a commercially available reagent. *Journal of Lipid Research* 30, 627–630.

Fagerlund, A., Shanks, D., Sunnerheim, K., Engman, L. and Frisell, H. (2003) Protective effects of synthetic and naturally occurring antioxidants in pulp products. *Nordic Pulp and Paper Research Journal* 18, 176–181.

Fernandez-Lopez, J., Zhi, N., Aleson-Carbonell, L., Perez-Alvarez, J.A. and Kuri, V. (2005) Antioxidant and antibacterial activities of natural extracts: application in beef meatballs. *Meat Science* 69, 371–380.

Finkel, T. and Holbrook, N.J. (2000) Oxidants, oxidative stress and the biology of ageing. *Nature* 408, 239–247.

Frankel, E.N. and Meyer, A.S. (2000) The problems of using one-dimensional methods to evaluate multifunctional food and biological antioxidants. *Journal of Science of Food and Agriculture* 80, 1925–1941.

Gordon, M.H. (1990) The mechanism of antioxidant action *in vitro*. In: Hudson, B.J.F. (ed.) *Food Antioxidants*. Elsevier, London/New York, pp. 1–18.

Guo, J.T., Lee, H.L., Chiang, S.H., Lin, F.I. and Chang, C.Y. (2001) Antioxidant properties of the extracts from different parts of broccoli in Taiwan. *Journal of Food and Drug Analysis* 9, 96–101.

Guo, Q., Zhao, B., Shen, S., Hou, J. and Xin, W. (1999) ESR study on the structure antioxidant activity relationship of tea flavon-3-ols and their epimers. *Biochimica et Biophysica Acta* 1427, 13–23.

Gupta, S. and Prakash, J. (2009) Studies on Indian green leafy vegetables for their antioxidant activity. *Plant Foods and Human Nutrition* 64, 39–45.

Halliwell, B. (1990) How to characterize a biological antioxidant. *Free Radical Research Communications* 9, 1–32.

Hamilton, R.J., Kalu, C., Prisk, E., Padley, F.B. and Pierce, H. (1997) Chemistry of free radicals in lipids. *Food Chemistry* 60, 193–199.

Herrmann, K. (1976) Flavonols and flavones in food plants: a review. *Journal of Food Technology* 11, 433–448.

Hua-Ming, C., Muramoto, K., Yamauchi, F. and Nokihara, K. (1996) Antioxidant activity of designed peptides based on the antioxidative peptide isolated from digests of a soybean protein. *Journal of Agricultural and Food Chemistry* 44, 2619–2623.

Huang, D., Ou, B. and Prior, R.L. (2005) The chemistry behind antioxidant capacity assays. *Journal of Agricultural and Food Chemistry* 53, 1841–1856.

Jayaprakash, G.K. and Patil, B.S. (2007) *In vitro* evaluation of the antioxidants activities in fruit extracts from citron and blood orange. *Food Chemistry* 101, 410–418.

Jayaprakash, G.K., Singh, R.P. and Sakariah, K.K. (2001) Antioxidant activity of grape seed (*Vitis vinifera*) extracts on peroxidation models *in vitro*. *Food Chemistry* 73, 285–290.

Jocelyn, P.C. (1972) *Biochemistry of SH Group*. Academic Press, London, p. 10.

Kahkonen, M.P., Hopia, A.I., Vuorela, H.J., Rauha, J.-P., Pihlaja, K., *et al.* (1999) Antioxidant activity of plant extracts containing phenolic compounds. *Journal of Agricultural and Food Chemistry* 47, 3954–3962.

Kahl, R. and Kappus, H. (1993) Antioxidantien BHA and BHT in vergleich mit daen naturlichen antioxidants vitamin E. *Zeitschrift für Lebensmittel-Untersuchung und Forschung* 196, 329–338.

Kakkar, P., Das, B. and Viswanathan, P.N. (1984) A modified spectrophotometric assay of super-oxide dismutase. *Indian Journal of Biochemistry and Biophysics* 21, 131–132.

Katalinic, V., Milos, M. and Jukic, M. (2006) Screening of 70 medicinal plant extracts for antioxidant capacity and total phenols. *Food Chemistry* 94, 550–557.

Katsube, T., Tabata, H., Ohta, Y., Yamasaka, Y., Annuurad, E., *et al.* (2004) Screening for antioxidant activity in edible plant products: comparison of low density lipoprotein oxidation assay, DPPH radical scavenging assay and Folin-Ciocalteau assay. *Journal of the Agricultural and Food Chemistry* 52, 2391–2396.

Kaur, C. and Kapoor, H.C. (2002) Antioxidant activity and total phenolic content of some of Asian vegetables. *International Journal of Food Science and Technology* 37, 153–161.

Kim, D.O. and Lee, C.Y. (2004) Comprehensive study on vitamin C equivalent antioxidant capacity (VEAC) of various polyphenolics in scavenging a free radical and its structural relationship. *Critical Reviews in Food Science and Nutrition* 44, 253–273.

Kim, S., Jeong, S., Park, W., Nam, K.C., Ahn, D.U., *et al.* (2006) Effect of heating conditions on grape seeds on the antioxidant activity of grape seed extracts. *Food Chemistry* 97, 472–479.

Koleva, I.I., van Beek, T.A., Linseen, J.P.H., Groot, A.D. and Evstatieva, L.N. (2002) Screening of plant extracts for antioxidant activity: a comparative study on three testing methods. *Phytochemical Analysis* 13, 8–17.

Kratchanova, M., Denev, P., Ciz, M., Lojek, A. and Mihailov, A. (2010) Evaluation of antioxidant activity of medicinal plants containing polyphenol compounds. Comparison of two extraction systems. *Acta Biochimica Polonica* 57, 229–234.

Kuppusamy, U.R., Indran, M. and Balraj, B.R.S. (2002) Antioxidant effects of local fruits and vegetable extracts. *Journal of Tropical Medicinal Plants* 3, 47–53.

Liu, H., Qiu, N., Ding, H. and Yao, R. (2008) Polyphenol contents and antioxidant capacity of 68 Chinese herbals suitable for medical or food uses. *Food Research International* 41, 363–370.

Longhi, G.J., Perez, E., Lima de, J.J. and Candido, L.M.B. (2011) *In vitro* evaluation of *Mucuna pruriens* (L.) DC. antioxidant activity. *Brazilian Journal of Pharmaceutical Sciences* 47, 535–544.

McCord, J. and Fridovich, I. (1969) Superoxide dismutase, an enzymic function for erythorocuprin. *Journal of Biological Chemistry* 244, 6409–6055.

Malmstrom, J., Engman, L., Bellander, M., Jacobsson, K., Stenberg, B., *et al.* (1998) Stabilization of PACREL[R] by organotellurium compound. *Journal of Applied Polymer Science* 70, 449–456.

Milic, B.L., Djilas, S.M. and Canadanovic-Brunet, J.M. (1998) Antioxidant activity of phenolic compounds on the metal-ion breakdown of lipid peroxidation system. *Food Chemistry* 61, 443–447.

Miller, D.M., Buettner, G.R. and Aust, S.D. (1990). Transition metals: catalysts of "autoxidation" reactions. *Free Radical Biology and Medicine* 8, 95–108.

Miller, N.J. and Rice-Evans, C.A. (1994a) Total antioxidant status in plasma and body fluids. *Methods in Enzymology* 234, 279–293.

Miller, N.J. and Rice-Evans, C.A. (1994b) Spectrophotometric determination of antioxidant activity. *Redox Report* 2, 161–171.

Moon, J.-K. and Shibamoto, T. (2009) Antioxidant assays for plant and food components. *Journal of Agricultural and Food Chemistry* 57, 1655–1666.

Niki, E., Tsuchihashi, H., Noguchi, N. and Gotoh, N. (1995) Interaction among Vitamin C, vitamin E and β-carotene. *American Journal of Clinical Nutrition* 62, 1322–1326.

Noda, Y., Anzai-Kmori, A., Kohono, M., Shimnei, M. and Packer, L. (1997). Hydroxyl and superoxide anion radical scavenging activities of natural source antioxidants using the computerized JES-FR30 ESR spectromoter system. *Biochemistry and Molecular Biology International* 42, 35–44.

Ohkawa, H., Onishi, N. and Yagi, K. (1979) Assay for lipid peroxidation in animal tissue by thiobarbituric acid reaction. *Analytical Biochemistry* 95, 351–358.

Osawa, T. and Namiki, M. (1981) A novel type of antioxidant isolated from leaf wax of Eucalyptus leaves. *Agricultural and Biological Chemistry* 45, 735–739.

Osawa, T., Yoshida, A., Kawakishi, S., Yamashita, K. and Ochi, H. (1995) Protective role of dietary antioxidants in oxidative stress. In: Cutler, R.G., Packer, L., Bertram, J. and Mori, A. (eds) *Oxidative Stress and Aging*. Birkhauser, Basel, Switzerland, pp. 367–377.

Ou, B., Huang, D., Hampsch-Woodill, M., Flanagan, J.A. and Deemer, E.K. (2002) Analysis of antioxidant activities of common vegetables employing oxygen radical absorbance capacity (ORAC) and ferric reducing antioxidant power (FRAP) assays: a comparative study. *Journal of Agricultural and Food Chemistry* 50, 3122–3128.

Oyaizu, M. (1986) Studies on product of browning reaction produced from glucose amine. *Japan Journal of Nutrition* 44, 307–315.

Ozcelik, B., Lee, J.H. and Min, D.B. (2003) Effects of light, oxygen and pH on the absorbance of 2,2 diphenyl-1-picrylhydrazyl. *Journal of Food Science* 68, 487–490.

Pellegrini, N., Serafini, M., Colombi, B., Del Rio, D., Salvatore, S., *et al.* (2003) Total antioxidant capacity of plant foods, beverages and oil consumed in Italy assessed by three different *in vitro* assays. *The Journal of Nutrition* 133, 2812–2819.

Prieto, R.L., Pineda, M. and Aguilar, M. (1999) Spectrophotometric quantitation of antioxidant capacity through the formation of phosphomolybdenum complex: specific application to the determination of vitamin E. *Analytical Biochemistry* 269, 337–341.

Prior, R.L. and Cao, G. (2000) Antioxidant phytochemicals in fruits and vegetables – diet and health implications. *Horticulture Science* 35, 588–592.

Ratty, A.K. and Das, N.P. (1988) Effects of flavonoids on non-enzymic lipid peroxidation: structure activity relationships. *Biomedical Medicine and Metabolic Biology* 39, 69–79.

Re, R., Pellegrini, N., Proteggente, A., Pannala, A., Yang, M., *et al.* (1999) Antioxidant activity applying an improved ABTS radical cation decolorization assay. *Free Radical Biology and Medicine* 26, 1231–1237.

Ren, W., Qiao, Z., Wang, H., Zhu, L. and Zhang, L. (2003) Flavonoids: promising anticancer agents. *Medicinal Research Reviews* 23, 519–534.

Rice-Evans, C.A., Miller, N.J. and Paganga, G. (1997a) Structure antioxidant activity relationships of flavonoids and phenolic acids. *Free Radical Biology & Medicine* 20, 933–956.

Rice-Evans, C.A., Miller, N.J. and Paganga, G. (1997b) Antioxidant properties of phenolic compounds. *Trends in Plant Science* 4, 304–309.

Robak, J. and Gryglewski, R.J. (1988) Flavonoids are scavengers of superoxide anions. *Biochemical Pharmacology* 37, 837–841.

Robards, K., Prenzler, P.D., Tucker, G., Swatsitang, P. and Glover, W. (1999) Phenolic compounds and their role in oxidative processes in fruits. *Food Chemistry* 66, 401–436.

Saito, S., Okamoto, Y. and Kawabata, J. (2004) Effect of alcoholic solvents on antiradical abilities of protocatechuic acid and its alkyl esters. *Bioscience, Biotechnology and Biochemistry* 68, 1221–1227.

Sasaki, Y.F., Kawaguchi, S., Kamaya, A., Ohshita, M., Kabasawa, K., *et al.* (2002) The comet assay with eight mouse organs: results with 39 currently used food additives. *Mutation Research – Genetic Toxicology and Environmental Mutagenesis* 519, 103–109.

Scott, G. (1985) A review of recent developments in the mechanisms of antifatigue agents. *Rubber Chemistry and Technology* 58, 289–293.

Serrano, J., Goni, I. and Saura-Calixto, F. (2007) Food antioxidant capacity determined by chemical methods may underestimate the physiological antioxidant capacity. *Food Research International* 40, 15–21.

Shahidi, F. (1997) *Natural Antioxidants: Chemistry, Health Effects and Applications*. AOCS Press, Champaign, Illinois, p. 5.

Sharma, O.P. and Bhatt, T.K. (2009) DPPH antioxidant assay revisited. *Food Chemistry* 113, 1202–1205.

Shizuka, S. and Kawabata, J. (2005) Effects of electron withdrawing substituents on DPPH radical scavenging reactions of protocatechuic acid and its analogues in alcoholic solvents. *Tetrahedron* 61, 8105–8108.

Sies, H. (1991) Oxidative stress: introduction. In: Sies, H. (ed.) *Oxidative Stress: Oxidants and Antioxidants*. Academic Press, London, pp. XV–XXII.

Singhal, P.K., Kapur, N., Dhillon, K.S., Beamish, R.E. and Dhalla, N.S. (1982) Role of free radicals in catecholamine cardiomyopathy. *Canadian Journal of Physiology and Pharmacology* 60, 1390–1397.

Sloane, H.J. and William, S.G. (1977) Spectrophotometric accuracy, linearity and adherence to Beer's law. *Applied Spectroscopy* 31, 25–30.

Spigno, G. and De Faveri, D.M. (2007) Antioxidants from grape stalks and marc: influence of extraction procedure on yield, purity and antioxidant power of the extracts. *Journal of Food Engineering* 78, 793–801.

Sroka, Z. and Cisowski, W. (2003) Hydrogen peroxide scavenging, antioxidant and antiradical activity of some phenolic acids. *Food and Chemical Toxicology* 41, 753–758.

Suja, K.P., Jayalekshmy, A. and Arumughan, C. (2005) Antioxidant activity of sesame cake extract. *Food Chemistry* 91, 213–219.

Sun, T. and Tanumihardjo, S.A. (2007) An integrated approach to compare food antioxidant capacity. *Journal of Food Science* 72, R159–R165.

Sun, T., Simon, P.W. and Tanumihardjo, S. (2009) Antioxidant phytochemicals and antioxidant capacity of biofortified carrots (*Daucus carota* L.) of various colors. *Journal of Agricultural and Food Chemistry* 57, 4142–4147.

Tabart, J., Kevers, C., Pincemail, J., Defraigne, J.O. and Dommes, J. (2009) Comparative antioxidant capacities of phenolic compounds measured by various tests. *Food Chemistry* 113, 1226–1233.

Tanaka, M., Kuei, C.W., Nagashima, Y. and Taguchi, T. (1988) Application of antioxidative maillard reaction products from histidine and glucose to sardine products. *Nippon Suisan Gakkaishi* 54, 1409–1414.

Tapiero, H., Tew, K.D., Ba, G.N. and Mathe, G. (2002) Polyphenol: do they play a role in the prevention of human pathologies? *Biomedical Pharmacotherapy* 56, 200–207.

Tripathi, Y.B. and Upadhyay, A.K. (2002) Effect of the alcohol extract of the seeds of *Mucuna pruriens* on free radicals and oxidative stress in albino rats. *Phytotherapy Research* 16, 534–538.

Viscidi, K.A., Dougherty, M.P., Briggs, J. and Camire, M.E. (2004) Complex phenolic compounds reduce lipid oxidation in extruded oat cereals. *Lebensmitt Wiss Technology* 37, 789–796.

Wargovich, M.J. (2000) Anticancer properties of fruits and vegetables. *Horticulture Science* 35, 573–575.

Wettasinghe, M. and Shahidi, F. (2002) Iron (II) chelation activity of extracts borage and evening primrose meals. *Food Research International* 35, 65–71.

Wickens, A.P. (2001) Ageing and the free radical theory. *Respiration Physiology* 128, 379–391.

Wojdyło, A., Oszmiański, J. and Czemerys, R. (2007) Antioxidant activity and phenolic compounds in 32 selected herbs. *Food Chemistry* 105, 940–949.

Wong, C.C., Li, H.B., Cheng, K.W. and Chen, F. (2006) A systematic survey of antioxidant activity of 30 Chinese medicinal plants using the ferric reducing antioxidant power assay. *Food Chemistry* 97, 705–741.

Wood, J.L. (1970) Biochemistry of mercapturic acid formation. In: Fishman, W.H. (ed.) *Metabolic Conjugation and Metabolic Hydrolysis*, Vol. II. Academic Press, New York, pp. 261–299.

Wootton-Bearda, P.C., Morana, A. and Ryan, L. (2010) Stability of total antioxidant capacity and total polyphenol content of 23 commercially available vegetable juices before and after *in vitro* digestion measured by FRAP, DPPH, ABTS and Folin-Ciocalteu methods. *Food Research International* 44, 3135–3148.

Wu, X., Beecher, G.R., Holden, J.M., Haytowitz, D.B., Gebhardt, S.E., *et al.* (2004) Lipophilic and hydrophilic antioxidant capacities of common foods in the United States. *Journal of Agricultural and Food Chemistry* 52, 4026–4037.

Yamaguchi, T., Takamura, H., Matoba, T. and Terao, J. (1998) HPLC method for evaluation of the free radical-scavenging activity of foods by using 1,1-diphenyl-2-picrylhydrazyl. *Bioscience, Biotechnology and Biochemistry* 62, 1201–1204.

Zulueta, A., Esteve, M.J. and Frigola, A. (2009) ORAC and TEAC assays comparison to measure the antioxidant capacity of food products. *Food Chemistry* 114, 310–316.

Index

Note: Page numbers in *italics* represent tables; those in **bold** represent figures.